STA'
USI

DAVID M. LEVINE
MARK L. BERENSON
DAVID STEPHAN
0-13-462912-4

January 1997

Dear Professor,

We have recently finished a new text, **Statistics for Managers Using Microsoft® Excel**, about which we are very excited. It is the first business statistics text to so thoroughly integrate the Microsoft Excel spreadsheet application that every statistical topic throughout the entire text is supported by Excel in explanation, example, and prepared workbooks on disk. We have done this in the belief that it is important for those of us in academia to provide improved linkage between the business world and both what and how we are teaching in statistics.

Statistics for Managers Using Microsoft® Excel is being printed now and will be available in late February 1997. In the meantime, Prentice Hall has put together this package of materials for your review. It includes the following:

- The Preface
- A detailed Table of Contents
- Three sample chapters:
  Chapter 1: Supplement: Introduction to Microsoft Excel
  Chapter 2: Presenting Data in Tables and Charts
  Chapter 11: Simple Linear Regression
- A list of statistical topics in the text supported by Excel
- A diskette containing all the data files and Excel workbooks corresponding to the sample chapters.

We welcome your questions or comments regarding the use of Excel in teaching statistics, how we have done so in our own classes at Baruch College, and our efforts here. Please feel free to contact us at the address and e-mail address below, or through Patrick Lynch at Prentice Hall (201-236-7644, patrick_lynch@prenhall.com); a complete set of page proofs is also available through Patrick.

Sincerely,

David M. Levine   Mark L. Berenson   David Stephan

Contact us at:
Baruch College
17 Lexington Ave.
New York, NY 10010
e-mail: dmlbb@cunyvm.cuny.edu

PRENTICE HALL BUSINESS PUBLISHING

STATISTICS FOR MANAGERS USING MICROSOFT® EXCEL

DAVID M. LEVINE
MARK L. BERENSON
DAVID STEPHAN
0-13-462912-4

WHO TO CONTACT FOR FURTHER INFORMATION

For further information, or to request an examination copy of this text, please contact your local Prentice Hall sales representative, or write to:

Patrick Lynch
Marketing Manager
Prentice Hall Business Publishing
College Marketing Department
One Lake Street
Upper Saddle River, NJ 07458
FAX: 1-201-236-7696
E-MAIL: patrick_lynch@prenhall.com

For Canadian orders, contact:

Prentice Hall – Canada
1870 Birchmount Road
Scarborough, Ontario M1P 2J7
Fax: 416-293-5646

For International orders and inquiries, contact your local Simon & Schuster International Group representative, or write to:

Simon & Schuster
200 Old Tappan Road
Old Tappan, NJ   07675 USA

You can also contact :

Faculty Services
1-800-526-0485

When writing to us, please include the book's title and ISBN.

Visit us on the World Wide Web at:

http://www.prenhall.com/phbusiness

PRENTICE HALL BUSINESS PUBLISHING

Integrating Excel into your business statistics course...

a video introduction!

Prentice Hall is pleased to offer a 10-minute video "symposium" in which David Levine and David Stephan discuss, instructor-to-instructor:

- Why use Excel?

- What are some of the advantages and disadvantages of using Excel in teaching business statistics?

- How can you integrate Excel into the classroom and curriculum?

Please contact your local Prentice Hall representative for copy of the video, or contact Patrick Lynch, Marketing Manager, at 201-236-7644, via e-mail at patrick_lynch@prenhall.com

Statistics for Managers Using Microsoft® Excel

DAVID M. LEVINE

Bernard M. Baruch College
City University of New York

MARK L. BERENSON

Bernard M. Baruch College
City University of New York

DAVID STEPHAN

Bernard M. Baruch College
City University of New York

 PRENTICE HALL, Upper Saddle River, New Jersey

Editor-in-Chief: Richard Wohl
Acquisitions Editor: Tom Tucker
Assistant Editor: Diane Peirano
Editorial Assistant: Audrey Regan
Marketing Manager: Patrick Lynch
Production Editor: Carol Lavis
Managing Editor: Katherine Evancie
Senior Manufacturing Supervisor: Paul Smolenski
Manufacturing Manager: Vincent Scelta
Senior Manager of Production and Technology: Lorraine Patsco
Electronic Page Make-up Artist: Christy Mahon
Electronic Art Supervisor: Warren Fischbach
Computer Artist: Steven Frim
Senior Designer: Suzanne Behnke
Design Director: Patricia Wosczyk
Interior and Cover Design: Suzanne Behnke
Cover Art: Kenny Beck

Copyright ©1997 by Prentice-Hall, Inc.
A Simon & Schuster Company
Upper Saddle River, New Jersey 07458

Printed in the United States of America
10   9   8   7   6   5   4   3   2   1

ISBN   0-13-632316-2

Prentice-Hall International (UK) Limited, *London*
Prentice-Hall of Australia Pty. Limited, *Sydney*
Prentice-Hall Canada, Inc., *Toronto*
Prentice-Hall Hispanoamericana, S.A., *Mexico*
Prentice-Hall of India Private Limited, *New Delhi*
Prentice-Hall of Japan, Inc., *Tokyo*
Simon & Schuster of Asia Pte. Ltd., *Singapore*
Editora Prentice-Hall do Brasil, Ltda., *Rio de Janeiro*

To our wives,
Marilyn L., Rhoda B., and Mary N.
and to our children,
Sharyn, Kathy, Lori, and Mark

Brief Contents

Contents

Chapter 3 Summarizing and Describing Numerical Data 119

Chapter 4   Basic Probability and Discrete Probability Distributions   171

Chapter 5   The Normal Distribution and Sampling Distributions   225

Chapter 8   Two-Sample and *c*-Sample Tests with Numerical Data   379

Preface

When planning this textbook, the authors focused on how desktop productivity tools, such as spreadsheet applications, have altered managers' decision-making processes. Whereas they once had to turn to a Management Information Systems Department or an Information Center to obtain customized summaries of corporate data, today an increasing number of managers use spreadsheet applications as the means to retrieve and analyze directly the data they need. In this context, employers now are beginning to desire, if not demand, that their college-educated, entry-level employees have more than just a cursory awareness of such tools as spreadsheet applications.

These changes, along with the realization that current spreadsheet applications can perform the type of analyses once done only by specialized statistical packages, have led us to develop *Statistics for Managers Using Microsoft Excel*. Our text contains the following features that distinguish it from the many other statistics texts available for business students (several of which have been written by two of us):

- Use of Microsoft Excel as a tool for statistical analysis throughout the text
- A streamlined version of topical coverage with sufficient breadth of coverage
- An enhanced managerial focus for statistical methods.

MAIN FEATURE: USE OF MICROSOFT EXCEL FOR STATISTICAL ANALYSIS THROUGHOUT THE TEXT

Statistics for Managers Using Microsoft Excel integrates the spreadsheet application Microsoft Excel throughout the entire text. This approach is fundamentally different from that of the many texts published and revised in the past twenty years. Since the advent of the computer revolution, statistics texts have struggled with the appropriate way to incorporate the use of statistical software packages. Most typically such packages as SAS, SPSS, and Minitab have been illustrated. A dilemma for faculty teaching this course has been how students could obtain access to (often through site licenses and student versions) the statistical software selected and how these packages could be used in the course. Often, students are not familiar with these packages prior to the statistics course, and only a limited number may use them in subsequent courses. Thus, students may view them as but one more hurdle to overcome in getting through the statistics course.

However, in the last several years, with the increasing functionality and power of spreadsheet applications, virtually all the kinds of statistical analysis taught in an introductory course are directly supported by the Microsoft Excel program (Version 5.0 or later), available for a variety of different systems including Windows 3.1, Windows 95, and Macintosh. In addition to its possible use in a statistics course, students typically learn the fundamentals of a spreadsheet application—either Microsoft Excel itself or a similar program—in an information systems course, and then use Excel in courses in accounting, finance, and other functional areas of business. Even if they are not familiar with Excel, they undoubtedly have heard of this software; and its use in the statistics course will give added relevancy to the course.

Because entering students' spreadsheet applications skills do vary, and because school and home computer facilities are sometimes limited, the demonstration of how Microsoft Excel can be incorporated into a statistics course must also vary. This text has been written for a number of situations, including the following:

1. **Statistical concepts with hands-on Excel development.** Instructors who wish to teach statistical concepts and the development of Excel-based solutions will find that the text includes detailed instructions for the design and implementation of Excel workbooks for each statistical topic discussed. These instructions are presented using a consistent developmental methodology that assists the student in developing their own solutions to statistical problems. (The methodology is summarized in Chapter 1S and is reflected in all of the Excel workbooks that are included on the diskette that accompanies this text.)

2. **Statistical concepts with Excel usage.** Instructors who wish to include Microsoft Excel in their courses but do not have the time or inclination to discuss specific steps of workbook development, can simply use the implemented examples on the diskette that accompanies this text, along with the portions of the Excel sections that analyze workbook results. The design of many of the diskette workbooks allows them to function as generalized templates into which different sets of data values can be entered and analyzed. (Each Excel section includes at least one workbook, and finished versions of all workbooks whose implementation is discussed in the text are included on the diskette that accompanies this text.)

3. **Statistical concepts with Excel exposure.** Where the instructor wishes to make students aware of the statistical applications of spreadsheets, but using Microsoft Excel directly is impractical, the results obtained from Excel for each statistical topic can be illustrated. (In these cases, the contents of the diskette can serve as the basis for optional assignments or student enrichment.)

Special Excel-related Features of the Text

1. **Excel orientation.** Because of the diverse computer backgrounds of incoming students, this text includes a comprehensive tutorial chapter, Introduction to Using Microsoft Excel (Chapter 1 Supplement), that assumes no previous experience using Excel or the windowing environments in which the program runs.

2. **Use of the workbook structure.** All the examples discussed in this text make full use of the Excel workbook feature to organize logically the data, calculations, and the results of a statistical analysis into different worksheets. In addition, all the workbooks included on the diskette that accompanies the text contain overview sheets that summarize the contents of the workbook and the relevant statistical concepts. The overview sheet that relates to the example discussed in Chapter 11 is illustrated in Figure P.1.Excel.

3. **Generalized instructions.** Instructions in the text for using Microsoft Excel will work equally well with the current versions of the program for the Macintosh, Windows 3.1, and Windows 95. (Details that are specific to particular windowing environments, such as the use of accelerator keys, have been avoided.)

4. **Alternative Excel approaches are contrasted.** Where appropriate, the text explores the differences between the use of Data Analysis tools and formula-based worksheets for statistical analysis.

5. **Coverage of an extensive set of Excel features.** The text includes a comprehensive discussion of the specialized functions, the PivotTable and Chart Wizards, the Text Import

	A	B	C	D	E	F	G
1			**Weekly Sales Analysis Workbook**				
2	Goal:	Using the Data Analysis tool for regression.					
3		**Cross-reference:**	Section 11.7				
4							
5	**Workbook Contents (three sheets):**						
6		**Overview Sheet**	Summary of this workbook.				
7		**Data Sheet**	Data for number of customers and weekly sales for a sample				
8			of 20 package delivery stores.				
9			Variable	Range	Values		
10			Store number	A2:A21			
11			Customers	B2:B21			
12			Sales	C2:C21	(weekly, in thousands of dollars)		
13		**Regression Sheet**	Results obtained from the Data Analysis Regression tool.				

FIGURE P.1.EXCEL   Overview sheet.

Wizard for importing text files into Microsoft Excel, the Data Analysis tools for statistical analysis, and the Scenario Manager as an aid in "What if" analyses.

In each chapter subsequent to Chapter 1, after a statistical topic has been covered, the use of Microsoft Excel as applied to the statistical topic is discussed in step-by-step detail. Each presentation of Excel material has the goal of ensuring that students will be able to use the standard features of Microsoft Excel to do what was just covered in the text. Thus, by the time the text is completed, students will have gained the necessary foundation in Excel to create their own workbooks to perform statistical and other types of analyses. An example of this detail can be seen in the discussion of the use of Excel to obtain descriptive statistics. In Sections 3.4.6 and 3.5.5 (pages 128–132 and 142–143), Excel functions are discussed; while in Section 3.7, the use of the Data Analysis tool is explained (pages 146–148). The output obtained from Excel is illustrated in Figure 3.7.Excel, which is duplicated below in Figure P.2.Excel.

	A	B
1	Sample Statistics for Pennsylv:	
2		
3	Tuition Variable	
4	Mean	8.3
5	Median	8.3
6	Mode	#N/A
7	Minimum	4.9
8	Maximum	11.7
9	Midrange	8.3
10	First Quartile rank	1.75
11	First Quartile	6.3
12	Third Quartile rank	5.25
13	Third Quartile	10.3
14	Midhinge	8.3

FIGURE P.2.EXCEL   Measures of central tendency obtained from Excel for out-of-state tuition rates for the six-school sample from Pennsylvania.

	A	B
	D1	
1	Calculating Normal Probabilities	
2		
3	Arithmetic Mean	75
4	Standard Deviation	6
5	Left Tail Probability	
6	First X Value	69
7	Z Value	-1
8	P(X<=69)	0.15865526
9	Right Tail Probability	
10	P(X>=69)	0.84134474
11	Interval Probability	
12	Second X value	81
13	P(X<=81)	0.84134474
14	P(69<X<81)	0.68268948
15	Finding a X Value	
16	Cumulative Percent	0.1
17	Z Value	-1.281550794
18	X Value	67.31069523

FIGURE P.3.EXCEL Design for computing normal probabilities illustrated for the individually trained factory worker (with $\mu = 75$ and $\sigma = 6$).

Numerous screen shots such as this are utilized throughout the text, and any statistical output obtained is explained in detail.

One of the advantages of using a spreadsheet application like Microsoft Excel is that it facilitates the use of "What if" analyses, which enable the student to explore the effect of changing data values. An example of such an analysis occurs in the computation of probabilities under the normal curve. Figure P.3.Excel, illustrated above, shows the use of Excel to find a probability under the normal curve.

For this situation, if we change the standard deviation to 10 we obtain the results shown in Figure P.4.Excel.

	A	B
	D1	
1	Calculating Normal Probabilities	
2		
3	Arithmetic Mean	75
4	Standard Deviation	10
5	Left Tail Probability	
6	First X Value	69
7	Z Value	-0.6
8	P(X<=69)	0.274253065
9	Right Tail Probability	
10	P(X>=69)	0.725746935
11	Interval Probability	
12	Second X value	81
13	P(X<=81)	0.725746935
14	P(69<X<81)	0.45149387
15	Finding a X Value	
16	Cumulative Percent	0.1
17	Z Value	-1.281550794
18	X Value	62.18449206

FIGURE P.4.EXCEL Design for computing normal probabilities illustrated for the individually trained factory worker (with $\mu = 75$ and $\sigma = 10$).

Numerous problems throughout the text include parts that ask the student to do this type of "What if" analysis. These problems are marked with an Excel icon next to the part that requires a "What if" analysis. This type of sensitivity analysis enhances understanding of the particular topic studied. To facilitate the use of numerous "What if" analyses, the Excel Scenario Manager feature is covered in the supplement to Chapter 1.

About the Diskette that Accompanies This Text

The diskette that accompanies this text includes Excel workbooks for all examples discussed in the text and all problems denoted with an Excel icon. The diskette also contains two visual basic for application modules. The first one (MOUSING.XLS) allows novices (and all others) to practice the mousing skills needed when using Microsoft Excel. The illustration below represents one portion of this workbook.

The second special module enables the user to generate a stem-and-leaf display from a set of values on a worksheet. All of these workbooks have been designed for and will load properly in Versions 5.0 or later of Microsoft Excel for the Macintosh, Windows 3.1, or Windows 95.

The diskette also includes files containing the data for the problems and examples in the text marked with a data disk icon ▣ in the margin. These files can be imported into Microsoft Excel using the Text Import Wizard as discussed in Chapter 1S. (Technical details about the diskette, including a complete list of the contents of the diskette is given in Appendix F.)

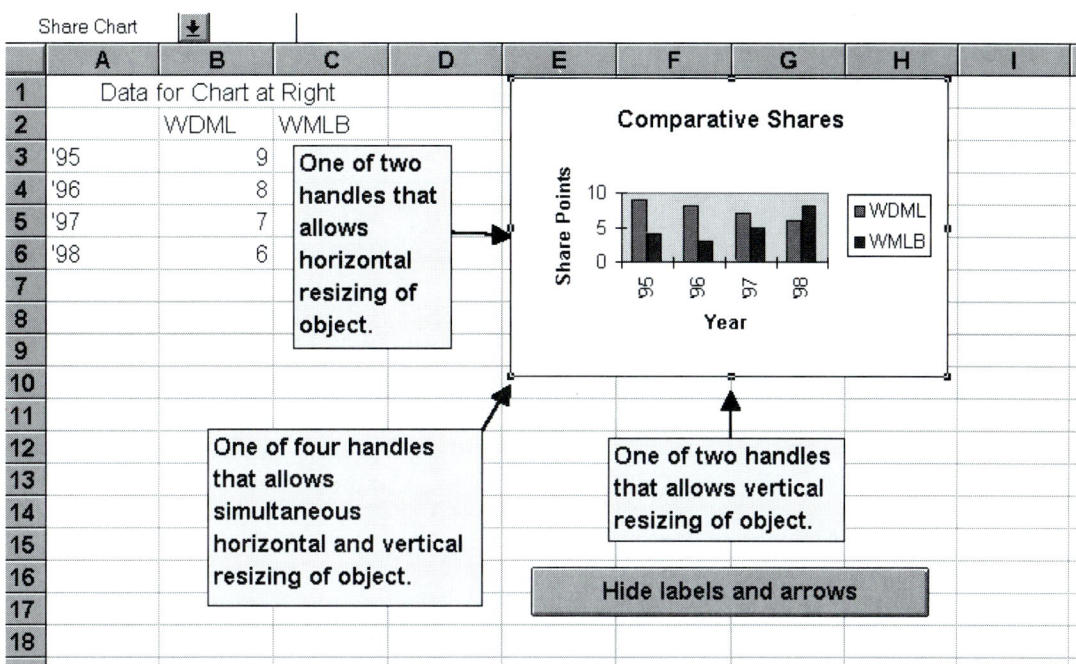

FIGURE P.5.EXCEL   Level 4 of mousing program.

Using Microsoft Excel

This text and the Excel workbook files on the diskette that accompanies this text have been designed for use with any of the following three versions of Microsoft Excel:

Version 5 (or 5.0c) for Windows 3.1

Version 7 for Windows 95

Version 5 for the Macintosh

These versions share the same command set and nearly identical dialog boxes for the operations related to the statistical analyses discussed in this text. (Section 1S.3, Using Microsoft Excel Dialog Boxes, details the differences between the three versions that occur when using dialog boxes for common computing tasks such as opening, saving, and printing files.)

As noted elsewhere, the instructions in the Excel sections of the text have been written to work equally well with any of the three versions. Although many of the instructions would also apply to Version 4 of Excel, this version should *not* be used for two reasons: Many problems are associated with the statistical routines in that earlier version and because Version 4 will not open any of the Excel workbooks included on the diskette that accompanies the text.

MAIN FEATURE: A STREAMLINED VERSION OF TOPICAL COVERAGE WITH SUFFICIENT BREADTH OF COVERAGE

The statistical coverage, as can be seen from the Table of Contents, provides more than sufficient breadth of coverage, although not the depth of coverage of other, more comprehensive texts such as Berenson and Levine's *Basic Business Statistics, 6th ed.* The first three chapters provide an introduction and cover tables, charts, and descriptive statistics. Chapters 4 and 5 discuss probability and probability distributions. Chapters 6–9 examine inference and hypothesis testing. Chapter 10 describes quality management. Chapters 11 and 12 talk about regression and multiple regression, and Chapter 13 ends the text with time series forecasting.

MAIN FEATURE: AN ENHANCED MANAGERIAL FOCUS FOR STATISTICAL METHODS

Although the statistical coverage in the text represents a more concise version of Berenson and Levine's *Basic Business Statistics, 6th ed.*, extensive rewriting provides for a more managerial focus. In numerous instances, notation has been simplified, and formulas such as Equation (3.7) below are written in words as well as statistical symbols.

> The standard deviation is the square root of the sum of the squared differences around the arithmetic mean divided by the sample size minus 1.
>
> $$S = \sqrt{\frac{\sum_{i=1}^{n}\left(X_i - \overline{X}\right)^2}{n-1}}$$
>
> (3.7)

Many of the important pedagogical features of Berenson and Levine, *Basic Business Statistics, 6th ed.* have been retained, including:

- Case Studies
- Thought-provoking action and "light bulb" problems

- Chapter-ending summary flow charts
- Team projects
- Discussion of ethical issues
- Extensive end-of-section and end-of-chapter problems
- Extensive use of real data throughout the text
- Listing of Key Terms

About the World Wide Web Icon

This text has a home page on the World Wide Web **WWW** with an address of http://www.prenhall.com/phbusiness. This home page includes a variety of information, including:

- Alternative course outlines
- Teaching tips
- Diskette additions and updates
- Links to other sites containing data appropriate for statistics courses
- Microsoft Excel–related sites

ACKNOWLEDGMENTS

We are extremely grateful to the many organizations and companies that allowed us to use their actual data for developing problems and examples throughout the text. We would like to thank *The New York Times,* Consumers Union (publisher of *Consumer Reports*), *U. S. News and World Report*, Moody's Investor Service (publishers of *Moody's Handbook of Common Stocks*), CEEPress Books, and Gale Research.

In addition, we would like to thank the Biometrika Trustees, American Cyanamid Company, the Rand Corporation, the American Society for Testing and Materials for their kind permission to publish various tables in Appendix E, and the American Statistical Association for its permission to publish diagrams from the *American Statistician.* Finally we are grateful to Professors George A. Johnson and Joanne Tokle of Idaho State University and Ed Conn, Mountain States Potato Company, for their kind permission to incorporate parts of their work as our Case Study F, "The Mountain States Potato Company."

A Note of Thanks

We would like to thank Kent S. Borowick, Baylor University; Ann Brandwein, Baruch College; Philip C. Fry, Boise State University; J. Morgan Jones, University of North Carolina; Stephen Reid, British Columbia Institute of Technology; Albert H. Segars, Clemson University; Richard Spinetto, University of Colorado; William E. Stein, Texas A & M University; Stanley D. Stephenson, Southwest Texas State University; Nancy C. Weida, Bucknell University; Peter H. Westfall, Texas Tech University; and Wayne Winston, Indiana University for their constructive comments during the writing of this textbook.

We would like to thank especially the editorial and production team at Prentice Hall for their assistance. In particular, we would like to thank our editor, Tom Tucker, without whose vision this text would not have become a reality. In addition, we would like to thank Richard

Wohl, Katherine Evancie, Joanne Jay, Diane Peirano, Audrey Regan, Carol Lavis, Christy Mahon, Lorraine Patsco, Paul Smolenski, Sue Behnke, and Patricia Wosczyk. Finally, we would like to thank our spouses and children for their patience, understanding, love, and assistance in making this book a reality. It is to them that we dedicate this book.

David M. Levine
Mark L. Berenson
David Stephan

Statistics
for Managers
Using
Microsoft®
Excel

Introduction to Using
MICROSOFT EXCEL

PART 1 ● Microsoft Excel Orientation

1S.1   WHAT ARE SPREADSHEET APPLICATION PROGRAMS?

In Section 1.2, when we discussed how software could help the manager use data to make decisions, we stated that a spreadsheet application program, Microsoft Excel, would be used as the software of choice in this text.

Spreadsheet application programs are the personal productivity programs best suited for the interactive manipulation of numerical data. These programs allow users to create electronic **spreadsheets,** or **worksheets,** which are rectangular arrays of (horizontal) rows and (vertical) columns into which entries are made.

Paper-based worksheets have long been used for the management and analysis of financial data by accountants. Not surprisingly, these professionals were among the first to understand the advantage of using *electronic* worksheets in which the effects of changing data could be immediately calculated and displayed.

Since their inception, spreadsheet programs have matured, gaining new functionality and growing more powerful over the years. In the case of Microsoft Excel, current versions of this program directly support most of the types of statistical analyses that are discussed in this text.

1S.1.1   Why Use Microsoft Excel?

Although other spreadsheet application programs can perform a variety of statistical analyses, Microsoft Excel's availability on a variety of personal computer operating systems and its dominant market share in business made it an easy choice for this text. Microsoft Excel also allows users to create electronic **workbooks**, collections of worksheets and other types of sheets. This feature can be used to develop solutions to statistical problems in a format that is the electronic equivalent of a typical (paper-based) management report. In this text, workbooks developed contain separate sheets that present the data and analysis for a problem as well as sheets that document and explain the spreadsheet methods used for the analysis. (The use of workbooks for this purpose is a pedagogical feature of this text.)

1S.1.2   Making the Most Effective Use of the Excel Sections

The Microsoft Excel sections in this text are designed to assist you in learning how to obtain and interpret pertinent statistics using features of Excel. Reviewing the worksheet designs and following the step-by-step instructions provided in the Excel sections while working with the Excel program is the most effective way of using this text. Through this active participation you will not only enhance your understanding of statistical concepts, but will also learn more about the specifics of both developing spreadsheet-based solutions and using the Microsoft Excel program.

The workbooks stored on the diskette that accompanies this text include and explain all the Excel examples from this text (see Appendix F for details). To enhance your learning, we suggest that you construct and save your own versions of the workbooks developed in the Microsoft Excel sections, following the instructions provided in those sections, using the workbooks on the accompanying diskette to double-check your work.

IS.1.3   Organization of This Supplement

In this supplement we first provide an orientation to the basic concepts necessary to operate any program, such as Microsoft Excel, that runs in a windowing environment such as Windows and the Macintosh operating system. We then survey the basic features and components of Microsoft Excel before we describe the planning methodology used to create all workbooks discussed in this text. The supplement concludes with the walkthrough of a solution to a simple statistical problem. This problem introduces you to the basic features of Microsoft Excel that enable you to use Excel as a tool for statistical analysis.

Computer novices should read all sections of this supplement; more experienced users may wish to proceed directly to the discussion of the planning methodology (Section 1S.6). Novice or not, upon finishing this supplement, you should be ready to use Microsoft Excel in a productive manner.

IS.2   USING WINDOWING ENVIRONMENTS

All versions of Microsoft Excel (version 5 for Windows 3.1, version 5.0 for the Mac OS, and version 7 for Windows 95) load and run in a **windowing environment**, a user interface in which **windows**, or frames, are used as containers to subdivide the screen. In windowing environments you accomplish tasks by pointing to and choosing on-screen objects that represent components or functions of your computer system. Typically, this pointing and choosing is done with a **mouse pointer,** an on-screen pointer that moves when you move a **mouse** or similar input device. Three mouse operations are frequently used in windowing environments and throughout our discussion of Microsoft Excel in this text and are defined as follows:

1. To **select** an on-screen object, you move the mouse pointer (by moving the mouse) directly over an object and then **click,** or press, the left mouse button (or click the single button, if your mouse contains only one button).

2. To **drag** an object, you move the mouse pointer over an object and hold down the left (or single) mouse button while moving the mouse.

3. To **double-click** an object, you move the mouse pointer directly over an object and click the left (or single) mouse button twice in rapid succession.

4. To **right-click** an object, you move the mouse pointer directly over an object and press the right mouse button. (If your mouse contains only one mouse button, right-clicking is the same as simultaneously holding down the control key and pressing the mouse button.)

> Mouse operations can be practiced using the MOUSING.XLS workbook, stored on the diskette that accompanies this text. To use this file, you will need to learn how to run the Microsoft Excel application and also how to open Excel workbook files (see Section 1S.3).

MOUSING.XLS

In a windowing environment, mouse operations are applied to a variety of on-screen objects. Typically, users select **icons**, graphics that represent a specific program application or document, in order to begin their work. (Selecting the Microsoft Excel program icon or an icon representing an Excel workbook as displayed in Figure 1S.1 would be typical ways to load and run Excel.)

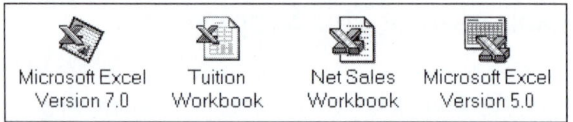

FIGURE 1S.1 **Microsoft Excel program icon and Excel workbook icon from the Microsoft Windows environment.**

Selecting on-screen objects in a windowing environment frequently triggers the display of **dialog boxes**, special windows that accept user input or report status information. In Microsoft Excel, dialog boxes appear to guide the user through such common computing tasks such as opening, saving, and printing files and can appear to help the user complete various spreadsheet and data analysis tasks. This section uses common task dialog boxes for opening and printing files to introduce and define objects found in all types of dialog boxes. Because common task dialog boxes differ according to the windowing environment being used by Excel, this section uses dialog boxes from both Excel version 5 for Windows 3.1 and version 7 for Windows 95 to illustrate the type of variation one would encounter when executing the same common task in different versions of Excel.

> Dialog boxes for the spreadsheet and data analysis functions of Excel are standardized across the versions for the Mac OS, Windows 3.1, and Windows 95, differing only in cosmetic details such as typeface used or the styling of the window frame. When illustrating such dialog boxes, the authors have chosen to use one illustration that will be recognizable to users of either of the two version 5's or version 7.

Figure 1S.2 shows the File Open dialog box as it appears in the Microsoft Excel version 5 for Windows 3.1. On the left-hand side of the dialog box, under the heading File Name is an **edit box,** an area on the screen into which you can directly type or edit a value, in this case the name of a file. Below the edit box is a linked **scrollable list box** that allows you to select

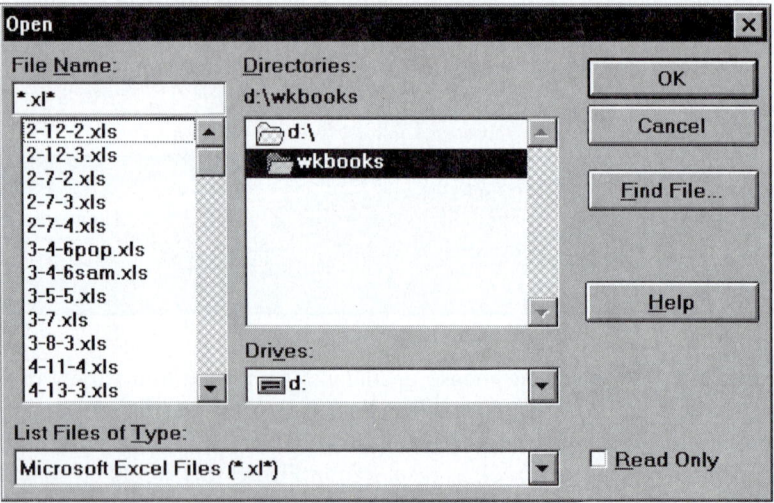

FIGURE 1S.2 **The File Open dialog box (version 5 for Windows 3.1).**

an item for the edit box by scrolling, or moving through, the list of choices. (Clicking the buttons on the **scroll bar,** immediately to the right of the list, is one way of scrolling through the list.)

In the center, under the label Directories, is another scrollable list box that appears without an accompanying edit box (a directory name must be picked from the list; it cannot be typed). Below the scrollable list boxes under the labels Drives and List Files of Type are two examples of a **drop-down list box.** These objects allow you to select an item from a non-scrollable list that appears when the drop-down button is clicked.

In the lower right is an example of a **check box,** a way of specifying an optional action. The optional action of this check box is to open a file as read only, meaning no changes can be saved to the file. (You may want to use this particular feature option when first opening and exploring a workbook file from the diskette that accompanies this text.)

Finally, on the right-hand side are four clickable buttons, two of which are common to many dialog boxes: the **OK button** (Open in version 7), that accepts all values, selections, and options as currently displayed in the dialog box and the **Cancel button,** that cancels the current operation (in this case, the opening of a file). Clicking these buttons will close the dialog box and cause it to disappear from the screen. (The version 7 File | Open dialog box, illustrated in Figure F.1 on page F–11, has a special properties button that can be used in conjunction with the Excel files on the diskette that accompanies this text. See Appendix F for details.)

Figure 1S.3 shows the File Print dialog box from version 7 of two additional objects commonly encountered, but which are not present in the File Open dialog boxes. In this File Print dialog box, notice that under the Print What and Page Range labels are two sets of **option buttons,** buttons that are used to present a set of mutually exclusive choices. Selecting an option button (also known as a radio button) always deselects, or clears, the other option buttons in the set, thereby preventing you from selecting two conflicting choices.

This dialog box also contains three sets of **spinner buttons** that appear to the right of the edit boxes associated with the copies:, from:, and to: labels. These pairs of buttons provide an alternate means of entering a value in the edit boxes by allowing you to increase or decrease the numeric value that appears in the associated edit box. (This version 7 dialog box, unlike the Print dialog boxes for the two version 5's, also allows the user to directly change the printer to be used.)

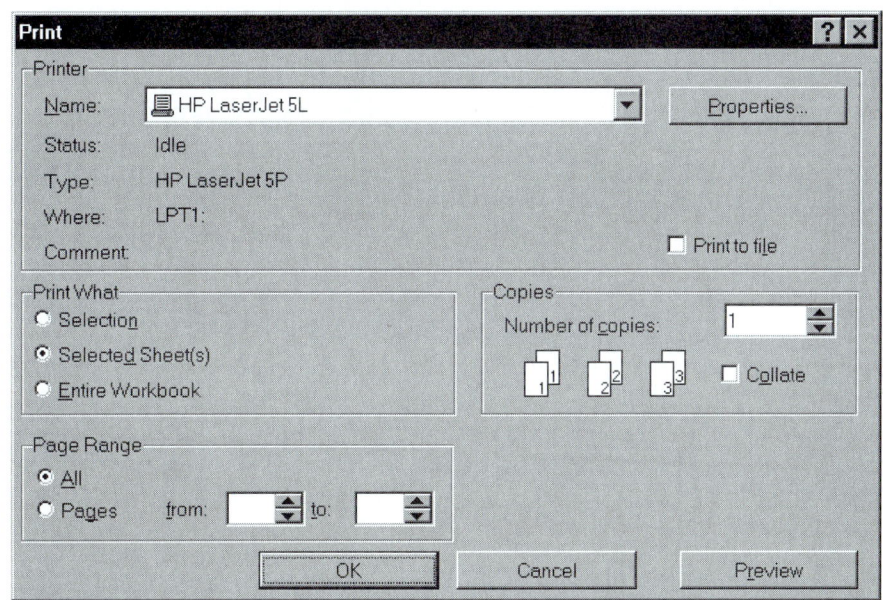

FIGURE 1S.3   The File Print dialog box (version 7 for Windows 95).

A sharp-eyed reader will have noticed that many menu choices, dialog box labels, and button captions contain an underlined character and that many menus include combination keystrokes, such as Crtl+P, as part of their menu choices. These are all examples of keyboard shortcuts that can be useful to a skilled typist. As they can vary from one windowing environment to another, such shortcuts are neither described nor used in this text.

IS.4 USING THE MICROSOFT EXCEL APPLICATION WINDOW

Double-clicking a Microsoft Excel program or workbook icon causes a windowing environment to load Excel and display the initial Excel application window. Figure 1S.4 shows this initial window for version 7, using the settings described later in this section. (The application windows for the other versions are similar).

This application window is divided into two parts, the lower of which displays the first sheet from an open (and new) workbook. It is into this lower area that you type entries in order to implement solutions for the problems you seek to solve. For now, focus on the upper part of the window which contains several **bars**, or rows of objects, that display important information or that perform the various functions of the program.

The first bar, the **title bar**, displays the name of the program and the name of the currently opened workbook ("Book1" in Figure 1S.4). The title bar also contains **buttons**, clickable graphics that simulate the operation of push buttons, that perform system tasks to control the display of the Excel window. (As these buttons vary across the different versions of Excel they are not described in detail here.)

The second bar, the **menu bar**, contains words and symbols that when selected cause Excel to execute some task or cause Excel to display a **pull-down menu**, a vertical list of fur-

FIGURE IS.4 Initial Microsoft Excel application window (version 7).

ther choices. For example, to display the File Open dialog box shown in Figure 1S.2, one would select the word File from the menu bar, and then from the subsequent pull-down menu select the word Open. Likewise, to display the File Print dialog box, one would again select File, but then select Print from the pull-down. (From this point on, we will abbreviate such selections by writing "File | Open" or "File | Print" using the vertical slash character | to separate menu choices.)

Immediately below the menu bar are two **toolbars**, containing buttons that perform common computing and worksheet formatting tasks. In Figure 1S.4, both the Standard and Formatting toolbars have been selected for viewing, and both have been snapped into position below the menu bar. (They can also appear as free-floating objects in their own mini-window.) Generally, the buttons on toolbars act as shortcuts, allowing the user to perform in a single mouse click what might otherwise require several menu selections. As you get proficient in using Excel, you may want to learn the shortcuts associated with these buttons.

Before beginning to use your copy of Microsoft Excel, you want to make your Excel windows appear as similar as is possible to Figure 1S.4 by following the boxed instructions that follow. (Again, as was noted in the discussion of dialog boxes in Section 1S.3, there may be some cosmetic differences to your display.) You may also want to verify that all of the spreadsheet features used in this text are properly installed and initialized in your copy of Excel.

To setup the Microsoft Excel application window as illustrated in Figure 1S.4, load the Excel program and issue the following commands:

1. **To display the two toolbars in the window.** Select View | Toolbars. The Toolbars dialog box appears. Select the Standard and Formatting check boxes and deselect all other toolbars on the list. Click the OK button.

2. **To display the formula and status bar and to standardize the display of the worksheet area.** Select Tools | Options. The Options dialog box appears containing two rows of labeled folder tabs. Select the View tab to make its view options visible if another tab's options are visible. (Normally, the view options will be visible when this dialog box is first displayed.) In the View Show groups, verify that the Formula bar and Status bar check boxes are selected. Under the Window Options group, make sure that Gridlines, Row & Column Headers, Horizontal scroll bar, Vertical scroll bar, and Sheet tabs check boxes are selected. Deselect the Formulas check box, if it has been selected. Click the OK button. (Other check boxes not mentioned can have the setting of your choice.)

To ensure that the Excel features used in this text are active and available in your copy of the program, load Excel and issue the following commands.

1. **To verify calculation and edit settings.** Select Tools | Options. In the Options dialog box that appears, select the Calculation tab. Verify that the Automatic calculation option button has been selected. Continue by selecting the Edit tab. Make sure that all check boxes *except* the Fixed Decimal check box have been selected. (Deselect the Fixed Decimal check box if necessary.) Click the OK box.

2. **To verify general Excel settings.** Select Tools | Options. In the Options dialog box that appears, select the General tab. Verify that the A1 option button has been selected. Change the value in the Sheets In New Workbook edit box to 4 if it is some other value. Verify that the Standard font is Arial and that the Size is 10. Use the drop-down list to select these values, if necessary. (Note: If Arial is not available on your system, use the font of your choice.) Click the OK button.

3. **To verify the installation of the Data Analysis Tools.** Select Tools. If Data Analysis appears as a menu choice, select it and verify that the dialog box that appears contains a scrollable list box in which the first item is "Anova: Single Factor" and the last choice is "z-Test: Two Samples for Means." (You will have to scroll the list to see the last choice.) Click the Cancel button. This verifies the installation of the tools.

If Data Analysis does not appear as a Tools menu choice, select Add-Ins…from the Tools menu. The Add-Ins dialog box appears. Select the Analysis ToolPak check box—if there is such a check box—from the "Add-Ins Available:" list. Click the OK button and exit Excel (Select File | Exit.). Reload Excel and follow the instructions in the preceding paragraph to verify installation of the tools.

If you cannot find an "Analysis ToolPak" check box in the "Add-Ins Available:" list, most likely the Data Analysis Tools component was not included when your copy of Excel was installed. (Consult with your technical support staff or review the chapter on installing Microsoft Excel in your Excel manual.)

IS.5 ORGANIZING EXCEL WORKBOOKS

Using spreadsheet application programs to develop accurate solutions to statistical problems requires careful planning and organization. Using these programs without first having a understanding of what is to be done and how it is to be done can cause us to produce worksheets—and in the case of Microsoft Excel, entire workbooks—that contain inaccurate or misleading results or that fail to clearly communicate the management information we seek.

To minimize such problems, we developed a standard format to help organize all the workbooks discussed in this text. This organization reflects the activities of analyzing the problem, designing a solution, implementing the solution, and verifying the solution that experienced users employ when developing any computer-based solution to a business problem. Understanding this format will help you make better use of the workbooks included on the diskette accompanying this text. (And because being able to properly organize solutions is such an important skill to master, a simple problem illustrating the use of our method is presented later in this supplement.)

Each of our workbooks contains an **Overview sheet** that serves as the table of contents for the workbook and that states the goals of the workbook, and the variables and the calculations needed to solve the problem. The overview sheet reflects our analysis of the problem and can be used by you as a guide to understanding the rest of the workbook.

The organization of each workbook also includes at least one **Data sheet** and one **Calculations sheet** that features the variables required by the problem and the calculations necessary to perform the statistical analyses required for the problem. The data and the calculations have been segregated in the workbooks to facilitate testing of the workbooks as well as to make it easier for you to modify the workbooks for use with other sets of data.

Where the results shown on a Calculations sheet are too cluttered for a clear and concise presentation, they have been reformatted onto an additional *Results sheet* that summarizes the solution to the problem. In a similar manner, when the programming techniques implemented to produce a solution are too complicated to easily summarize on the Overview sheet, they are explained separately in a *Behind the Scenes* sheet.

Many workbooks also contain additional sheets that have been created by Microsoft Excel as the result of using a feature of the program such as the Chart Wizard or the Data Analysis Tool. In addition, several workbooks contain "macro" procedures that have been placed in separate *module sheets* as required by the Excel program.

PART 2 ● Getting Started with Microsoft Excel

The next 11 sections guide you through all the steps of analyzing, developing, implementing, and testing that we used to develop the workbooks of this text. As you read these sections, you will also be introduced to the basic concepts and operating procedures necessary to construct workbooks in Microsoft Excel. (This knowledge is assumed in Excel discussions that appear later in this text.)

1S.6 DESIGNING SHEETS FOR A WORKBOOK

To illustrate the development methodology described in the previous section, suppose that we have been asked to explore the variability among the costs that would be incurred attending any one of six colleges in the state of Pennsylvania. Assume that we have been provided data for the three variables (school, tuition, and room and board) that are shown in Table 1S.1 (These data, from a sample drawn from a population of 90 Pennsylvania colleges, will also be used when we discuss descriptive statistics in Chapter 3.)

As the first step in developing a Microsoft Excel workbook solution, we decide to express the problem as "What are the maximum and minimum values for the tuition and

Table 1S.1 Tuition and room and board expenses for a sample of six Colleges in Pennsylvania.

SCHOOL	TUITION (IN $000)	ROOM AND BOARD (IN $000)
University of Pittsburgh	10.3	4.1
East Stroudsburg University	4.9	2.9
Geneva College	8.9	4.1
Drexel University	11.7	3.8
California Univ. of Penn.	6.3	2.9
Slippery Rock University	7.7	3.4

Table 1S.2 Initial design of the Data sheet for the Pennsylvania college costs analysis data ($000).

SCHOOL	TUITION	ROOM AND BOARD	TOTAL COST
xxxxx	xx.x	x.x	xxx.x
xxxxx	xx.x	x.x	xxx.x
.	.	.	.
.	.	.	.
.	.	.	.
xxxxx	xx.x	x.x	xxx.x

room and board variables in this sample?" We also decide to ask the same question of a total cost variable, which we will define as the sum of the tuition and room and board expenses for each school.

Continuing on, we determine that a listing of the maximum and minimum values for these three variables, along with the range (the difference between the maximum and the minimum value) from each of the three will be the results. (See Section 3.5 for a complete discussion of the range statistic.) For our design for the Data Sheet, we develop Table 1S.2, using x's and ellipses as place holders for actual data that will entered into the table. (We place a decimal point in the tuition, room and board, and total cost place holders to remind ourselves that the data were recorded to the nearest tenth in thousands of dollars.)

> ### A Good Practice: Data Sheet Design
>
> Note that the design of the data sheet reflects the arrangement of values as found in Table 1S.1 since this is the source of the data for our problem. Designing a data sheet in this manner is a good practice, one that will minimize your data entry errors later.

For this (relatively simple) problem, we decide that the Calculations sheet can serve as the result sheet as well. We produce Table 1S.3 as the initial design for this sheet; we will postpone the exact details of the sheet later in the development process.

Table 1S.3 Initial design of the Calculations sheet Pennsylvania college costs analysis.

	TUITION	ROOM AND BOARD	TOTAL COST
Maximum	xxxx	xxxx	xxxx
Minimum	xxxx	xxxx	xxxx
Range	xxxx	xxxx	xxxx

1S.7   DESIGNING THE DATA SHEET

Having completed our initial analysis and design of the sheets for this problem, we are now ready to implement them as worksheets in a Microsoft Excel workbook. Doing this will require mapping specific elements of the design such as the title and column headings to specific locations in a specific worksheet.

Recall that earlier in this chapter, worksheets were identified as rectangular arrays of (horizontal) rows and (vertical) columns into which entries are made. Mapping, therefore, requires associating the columns and lines (rows) of our initial designs with specific columns and rows in a worksheet. Using a standard notation for worksheets, we can label the columns of our initial design with letters and label our lines (rows) with numbers to identify into which **cells** (the intersections of the rows and columns) the entries should be made.

For our Data sheet, we decide that we should start with the column headings of Table 1S.1 and that the first column heading (School) should appear in cell A1; that is, it should be entered into the intersection of the first ("A") column and the first ("1") row. From this we quickly determine that the tuition and room and board headings should appear in cells B1 and C1, respectively, and that the values of each variable for the six colleges should appear in the next six rows (2–7). With these decisions made, we can revise Table 1S.2 to produce a design that can be directly implemented into a worksheet (see Table 1S.4). (Later in this supplement, the initial design for the Calculations sheet will be similarly revised.)

Table 1S.4 Revised design of the Data sheet for the Pennsylvania college costs analysis.

	A	B	C	D
1	School	Tuition	Room and Board	Total Cost
2	xxxxx	xx.x	x.x	xxx.x
3	xxxxx	xx.x	x.x	xxx.x
4	.	.	.	.
5	.	.	.	.
6	.	.	.	.
7	xxxxx	xx.x	x.x	xxx.x

Cells and Cell Addresses

In the preceding section, "A1" is an example of a **cell address.** Many times, a cell address in this column letter and row number format will be sufficient to specify the mapping of a cell entry. At other times—for example, when a design is implemented that references cells in other sheets or when certain features of Microsoft Excel are used—the address will need to include the name of the sheet. In such cases, the cell address is written in the form *sheetname!column-row* such as Data!A1 for the cell in the first column and row of the sheet named "Data." This form allows you to distinguish between two similarly located cells of two different sheets—for example, Data!A1 and Calculations!A1. (Should the name of the sheet include one or more spaces, then the sheet name needs to be enclosed in a pair of single quotation marks as in 'Sorted Data'!A1.)

Note that when a cell address does not contain the name of a sheet, the current sheet is implied. See discussion about formulas in Section 1S.9 to learn more about how this can minimize some typing when implementing a worksheet design.

1S.8 ENTERING VALUES INTO A DATA SHEET

Having specified cell addresses for the various parts of the Data worksheet in the previous section, we are now ready to enter values into the cells of the Data sheet. To do this, activate the Microsoft Excel application, and select File | New to create a new, blank worksheet window. Rename Sheet1 as Data, following the instructions given in Section 1S.5.

Select cell A1 by clicking its interior. A special border, the **cell highlight,** appears around the cell. This highlight indicates that cell A1 is now the **active cell**—that is, the cell into the next value to be typed will be entered. (Also note that A1, the address of the active cell, appears in the cell reference box.) Type the column heading School. As you type, notice that your keystrokes appear both in the edit box of the formula bar as well as in cell A1 itself. Press the Enter key (or click the check button to the left of the edit box) to complete the entry. (Users of keyboards that do not contain an Enter key, should press the Return key when instructions in this text call for pressing the Enter key.) If you made a mistake typing, you can use one of the methods described in the "Correcting Errors" box to correct your mistake.

Correcting Errors

At some point, as you enter text or numbers, you will probably make an entry that needs to be corrected. When you make a mistake while typing an entry, you can press the Escape key to cancel the entry, or you can use one of the following methods to edit your entry.

- Press the Backspace key to erase characters to the left of the cursor one character at a time.
- Press the Delete key to erase characters to the right of the edit cursor one character at a time.
- Replace the in-error text by clicking at the start of the error and dragging the mouse pointer over the rest of the error and then typing the replacement text.
- If you change your mind, you can undo your last edit by selecting the command: Edit | Undo. (There is also an Edit | Redo should you change your mind a second time and wish to keep the edit after all.)

As you complete this entry, notice that part of the heading extends into the next column. A way of overcoming this problem by having the columns automatically adjusted will be discussed in Section 1S.11 (Enhancing the Appearance of the Worksheet).

Continue by selecting cell B1 and entering the column heading Tuition. Then select cell C1 to enter the Room and Board column heading. Finally, select cell D1 and enter the column heading Total Cost.

Now that all column headings have been entered, we can begin to enter the values that will appear underneath them. We will type in values by columns, using the feature of the Enter (Return) key to automatically advance the cell highlight down one row after each entry. (If we wished to enter values by rows, we could end each entry by pressing the Tab key, which would advance the cell highlight one column to the right.)

Select cell A2 and type the name University of Pittsburgh and press the Enter (or Return) key. Then type East Stroudsburg University in cell A3 and press the Enter key. Now type Geneva College in cell A4, press Enter; type Drexel University in cell A5, press Enter; type California Univ. of Penn. in cell A6, press Enter; and finally type Slippery Rock University in cell A7 and press Enter.

Select cell B2 and enter the tuition values from Table 1S.1 into cells B2:B7. Select cell C2 and enter the room and board values data from Table 1S.1 into cells C2:C7. Because the total cost variable was defined (unlike the other variables), and not taken from our source table, we momentarily defer until Section 1S.10 an explanation of how to enter the data for this column.

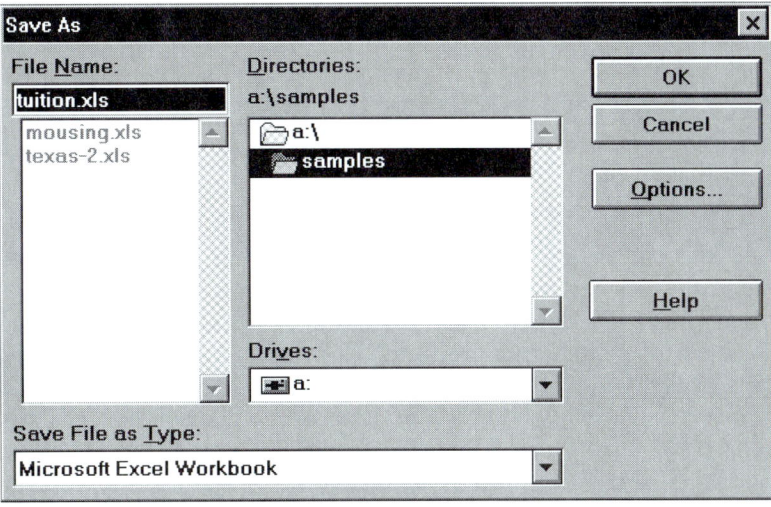

FIGURE 1S.5 The File Save As dialog box (version 5 for Windows 3.1).

Instead, having now entered all the data from our source table in the Data worksheet, we should **save,** or store a copy of our work on disk, before continuing. To save our work, select File | Save As. The File Save As dialog box which appears (see Figure 1S.5) will be similar to the File | Open dialog box for the version of Microsoft Excel being used. Specify the appropriate choices for drive, directory, and file name under which to store the workbook. (These choices vary based on the type of system used and whether you are saving your work to a diskette, local hard disk, or network disk. Ask your instructor if you are unsure as to the values you should choose.)

A Good Practice

After you have finished a single task, such as entering all the data from a source table, stop and save your work. Don't wait until you have finished all the tasks associated with developing a workbook. Unless otherwise directed by your instructor, you should always name your files using a name that *does not match* any of the files on the diskette that accompanies this text. This will prevent accidental alteration of the files on the diskette and avoid possible confusion.

1S.9 USING FORMULAS IN A DATA SHEET

Now that we have saved our work, we are ready to include the data for the total cost variable in cells D2:D7 of the Data worksheet. Recall that since this variable was not part of our original source table, we defined it in Section 1S.7 to be equal to the sum of the tuition and room and board expenses for each school. Therefore, one way to generate the total cost data would be to follow this definition and manually compute the values—for example, adding 10.3 and 4.1 to get 14.4 as the value for the total cost variable for the University of Pittsburgh.

Although it might be argued that for this very small and very simple problem, manual calculation would be the best method to generate the total cost data, it is generally wiser to have Microsoft Excel generate the values than to do it yourself. For Excel to generate these values, we will need to develop and enter **formulas,** or instructions to perform a calculation or some other task, in the appropriate cells of our Data worksheet (cells D2:D7 in this example).

Table 1S.5 Arithmetic Operators in Microsoft Excel

ARITHMETIC OPERATION	EXCEL OPERATOR
Addition	+
Subtraction	–
Multiplication	*
Division	/
Exponentiation (a number raised to a power)	^

To distinguish them from other types of cell entries, all formulas always begin with the = (equals sign) symbol. Creating formulas requires knowledge of the **operators,** or special symbols, used to express arithmetic operations. Operators used in formulas in this text are listed in Table 1S.5.

As our definition of the total cost variable calls for the addition of two quantities, we will use the + (plus sign) in our total cost formula, combining it with the cell addresses that contain the values we wish to add. In the case of calculating the total for the University of Pittsburgh, cells Data!B2 and Data!C2, which contain the tuition and room and board values for that school, are added together.

Assembling these pieces, we can form the formula =Data!B2+Data!C2 and enter it into cell D2. However, because we are entering a formula in the same sheet as the sheet to which it refers, we can write the formula using the shorthand notation =B2+C2, and Microsoft Excel will correctly interpret the addresses as referring to the current (Data) sheet.

Similar formulas can now be constructed and entered for the D column cells in rows 3–7. We adjust the original formula so as to refer to the tuition and room and board variables for the other five colleges. The formulas are as follows:

=B3+C3 for cell D3, the total cost variable for East Stroudsburg University

=B4+C4 for cell D4, the total cost variable for Geneva College

=B5+C5 for cell D5, the total cost variable for Drexel University

=B6+C6 for cell D6, the total cost variable for California Univ. of Penn.

=B7+C7 for cell D7, the total cost variable for Slippery Rock University

Again, note that the results of the formulas, and not the formulas themselves, are displayed in the worksheet.

1S.10 USING COPY TO FACILITATE THE ENTRY OF FORMULAS

As the construction and entry of similar formulas is a common task when developing workbooks, Microsoft Excel includes a copy shortcut feature that can facilitate this task.

To demonstrate this feature, let's pretend that we have not entered any of the formulas of the previous section. (If you have already entered those formulas, select cell D2 and press the Delete key to delete the formula in cell D2. Then do the same for cells D3–D7 to delete the formulas in those cells.) Begin by entering the first formula for the total cost variable, =B2+C2, into cell D2. Select cell D2 as shown in Figure 1S.6 Panel A. Then move the mouse pointer so that it is over the square handle in the lower right-hand corner of the cell highlight. The mouse pointer changes from an outlined plus sign to a plain plus sign (Panel B). Then drag the mouse pointer through the group of cells or **cell range** D3:D7 that will contain the similar formulas (see Panel C). Release the mouse button to complete the operation.

D2 | =B2+C2

	A	B	C	D
1	School	Tuition	Room and Board	Total Cost
2	University of Pittsburgh	10.3	4.1	14.4
3	East Stroudsburg University	4.9	2.9	
4	Geneva College	8.9	4.1	
5	Drexel University	11.7	3.8	
6	California Univ. of Penn.	6.3	2.9	
7	Slippery Rock University	7.7	3.4	
8				

D2 | =B2+C2

	A	B	C	D
1	School	Tuition	Room and Board	Total Cost
2	University of Pittsburgh	10.3	4.1	14.4
3	East Stroudsburg University	4.9	2.9	
4	Geneva College	8.9	4.1	
5	Drexel University	11.7	3.8	
6	California Univ. of Penn.	6.3	2.9	
7	Slippery Rock University	7.7	3.4	
8				

=B2+C2

	A	B	C	D
1	School	Tuition	Room and Board	Total Cost
2	University of Pittsburgh	10.3	4.1	14.4
3	East Stroudsburg University	4.9	2.9	
4	Geneva College	8.9	4.1	
5	Drexel University	11.7	3.8	
6	California Univ. of Penn.	6.3	2.9	
7	Slippery Rock University	7.7	3.4	
8				

FIGURE 1S.6 The copying operation in Microsoft Excel. Panels A, B, and C.

Microsoft Excel pastes formulas adjusted to refer to the proper cells, and not exact copies of the formula. In this example, since we were copying down a column, Excel adjusted the row portion of all addresses in the formulas, just as we did earlier when we manually entered those formulas in the previous section. This can be verified if we select any of the cells in range D3:D7 and examine the cell's contents in the formula bar area. (We need to select a cell and examine the formula bar area because, as noted above, cells containing formulas display the result of the formula and not the formula itself.)

With the inclusion of these formulas, all the entries to the Data worksheet have now been made. (Before continuing, this would also be another good place to save our workbook using the procedure described in the Good Practices box at the end of the Section.1S.8.)

1S.11 ENHANCING THE APPEARANCE OF A DATA SHEET

Now that we have completed all the entries of the Data worksheet, we can make changes to enhance the appearance of the Data worksheet. The first improvement involves adjusting the width of the columns so that all column headings and column values can be easily read. This can be done by moving the mouse pointer to the column headings and clicking on the column heads for columns A–D and issuing the command Format | Column | Autofit. Notice that the column widths have automatically adjusted to the widest entry in each column.

Another improvement involves changing the column headings from a plain format to bold-face. Perhaps the simplest way to do this is to first select the cell range to be reformatted (cells A3:D3) and then to click the B (boldface) button of the formatting toolbar. (To select the range A3:D3, first select cell A3 and then drag the mouse pointer across cells B3 through D3.) Then, with these three cells highlighted, click on the B (boldface) button on the formatting toolbar.

Another change we could consider is making the column headings appear centered over their columns. (Notice that they are currently left-justified in their columns.) If you wish to center the headings, select the range A3:D3 as before and click on the center button on the formatting toolbar. The column headings for columns A–D are now centered.

Having previously entered the numbers in columns B and C and the formulas in column D, you should observe that the numeric values in column B are not aligned on the decimal point. This problem may be fixed by first selecting the range B2:D7 and then clicking on the increase decimal point box on the formatting toolbar. You will now see that all values displayed have a single decimal point and are aligned on the decimal.

An additional improvement that is useful is centering a title over a range of columns. If we wanted to center a first-row worksheet title over a range of columns, we would do the following:

❶ Enter the title in the leftmost cell in the desired row.

❷ Select the range of cells in which the title is to be centered.

❸ Issue the command Format | Cells.

❹ Select the Alignment tab on the Cell Format dialog box.

❺ Select the center Across Selection button and click the OK button.

At this point, you should again save your work, using the procedure described in the Good Practices box at the end of the Section 1S.8.

1S.12   USING FUNCTIONS IN FORMULAS

In Section 1S.9, we used the plus sign arithmetic operator to construct our formula. We could have just as easily used the Sum **function,** one of many such pre-programmed instructions that can be used when solving a variety of common arithmetic, business, engineering, and statistical problems.

To use the Sum function, we would have typed the formula =SUM(B2:C2) into cell D2 instead of the formula =B2+C2. In the formula =SUM(B2:C2), the word SUM identifies the sum function, the pair of parentheses () bracket the cells of interest, and B2:C2 is the address of the cell range of interest, the cells whose values will be used by the function. (Since using the colon in this manner to express a range is so common in Excel, it will be the standard for indicating ranges in the rest of the text.)

Formulas that contain functions are otherwise identical to formulas that do not. This means that in this case we could individually type the formulas =SUM(B3:C3), =SUM (B4:C4), =SUM(B5:C5), =SUM(B6:C6), and =SUM(B7:C7), into cells D3:D7, respectively, or use the copy shortcut feature to place these formulas into the proper cells.

Earlier in this supplement we decided that one worksheet would serve both for the calculations as well as for the results for the college costs analysis, and we developed Table 1S.2 as the design for this worksheet. In this section we implement that sheet, illustrating the use of additional functions as we go along.

Recall that for our results, we wished to present the maximum and minimum values, along with the range (defined as the difference between the maximum and the minimum value), for the three variables of interest. We can use the MAX and MIN functions in formulas to produce the maximum and minimum values for the variables and use the subtraction operator (the minus sign) in formulas to produce the range values.

We decide to map the title for the sheet to cell B1, the column headings Tuition, Room and Board, and Total Cost to cells B3, C3, and D3, and the labels Maximum, Minimum, and Range to cells A4, A5, and A6, respectively. We revise our initial design, replacing the place holders with the formulas to be entered into the worksheet. Table 1S.6 shows our revision.

Table IS.6 Revised design of the Calculations sheet for the Pennsylvania college costs analysis.

	A	B	C	D
1		Pennsylvania College Costs Analysis		
2				
3		Tuition	Room and Board	Total Cost
4	Maximum	=MAX(DATA!B2:B7)	=MAX(DATA!C2:C7)	=MAX(DATA!D2:D7)
5	Minimum	=MIN(DATA!B2:B7)	=MIN(DATA!C2:C7)	=MIN(DATA!D2:D7)
6	Range	=B4 – B5	=C4 – C5	=D4 – D5

Note the inclusion of the worksheet name (Data) in the ranges used by the MAX and MIN formulas. This inclusion is necessary since we are referring to a range of cells on another sheet. Note that the range could also be entered as DATA!B2:DATA!B7, but we chose to use the same shorter way that the Excel program uses when it reports ranges in certain dialog boxes.

As for the range formulas, we could have constructed formulas in the form of =MAX(range) – MIN(range), for example, =MAX(DATA!B2:B7) – MIN(DATA!B2:B7) for the range of the tuition variable, but we chose the simpler alternative, which has the added advantage of more clearly communicating the relationships among the cells of this worksheet.

To implement this design, we would activate the Microsoft Excel application and open our Tuition workbook. We then would select the tab of a previously unused worksheet (e.g., Sheet2), rename the sheet "Calculations," and enter all the values and formulas in the proper cells.

A completed worksheet is shown in Figure 1S.7. After entering all values and formulas, you should review all entries for errors and test all calculations with simple numbers. Review

A1		School		
	A	**B**	**C**	**D**
1	School	Tuition	Room and Board	Total Cost
2	University of Pittsburgh	10.3	4.1	14.4
3	East Stroudsburg University	4.9	2.9	7.8
4	Geneva College	8.9	4.1	13
5	Drexel University	11.7	3.8	15.5
6	California Univ. of Penn.	6.3	2.9	9.2
7	Slippery Rock University	7.7	3.4	11.1

FIGURE IS.7   Completed Calculations sheet for the Pennsylvania college costs analysis.

Customizing Your Printouts

The File | Page Setup command offers numerous options that control the way your printed pages look. Selecting this command produces the Page Setup dialog box illustrated in Figure 1S.8.

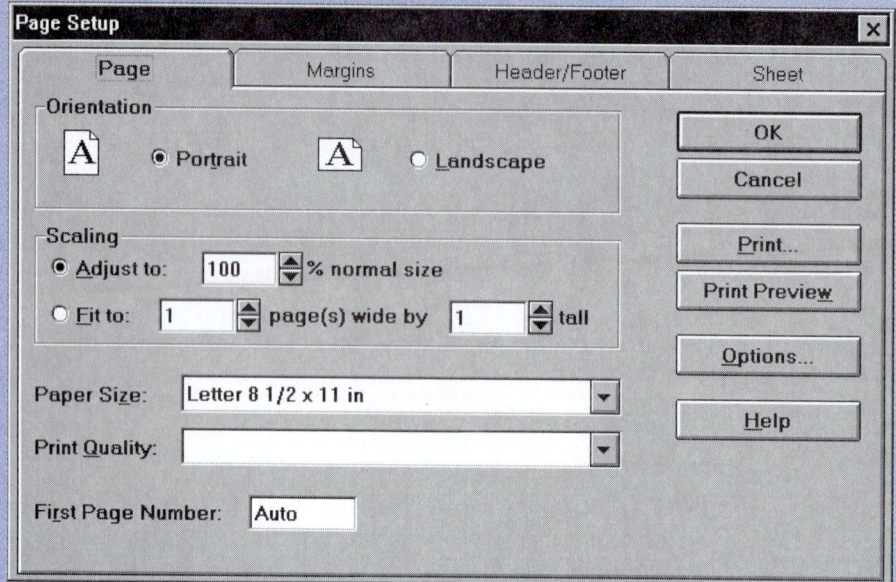

FIGURE 1S.8   The Page Setup dialog box.

The Page Setup dialog box contains tabs labeled Page, Margins, Header/Footer, and Sheet. Some of the choices available from these tabs are useful to know about.

● **Page tab choices.**   The **Orientation option buttons** control whether sheets from your workbook are printed vertically (Portrait) or horizontally (Landscape). Landscape orientation is particularly useful for worksheets that have a greater width than length. The **Scaling option buttons** allow you either to reduce or enlarge the printed size of a worksheet (Adjust option) or to specify the number of pages to fit on one printed page (Fit option).

● **Margins tab choices.**   Choices on this tab control the size of all margins on the printed page as well as whether the printed worksheets are centered on a page.

● **Header/Footer tab choices.**   Choices here allow you to choose from many different styles of headers and footers that can be added to your printouts. You can also construct your own custom headers and footers.

● **Sheet Tab choices.**   Most useful among the many Sheet choices are the Print check boxes that allow you to specify whether gridlines and row and column headings are included in the printouts of your worksheets. There is also a check box to specify draft quality printout; this is particularly useful when using older and slower types of printers.

of all formulas can be greatly aided by selecting the Tools I Options command and then, when the Options dialog box appears, selecting the Formulas checkbox of the View tab. This causes Excel to display the formulas—and not their results—on-screen. After this adjustment, you may need to issue the Format | Column | Autofit command discussed in Section 1S.11 to view all the formulas clearly. (You can then print the worksheet for later review by selecting the File | Print command and responding appropriately to the choices in the dialog box that appears. See the discussion of the File Print dialog box in Section 1S.3).

1S.14 INTRODUCTION TO WHAT IF ANALYSIS

As mentioned earlier in this supplement, one of the advantages of using spreadsheet applications is that the results of changing data can be immediately calculated and displayed. This dynamic nature of spreadsheet applications is especially valuable in statistical analysis, where it can be used to explore the sensitivity of the effect of changes in some portion of a data set on the results obtained. Frequently, we change the values of one variable and note the new results, a process that is commonly called "what if analysis."

With our completed and tested Tuition workbook, we could ask "What would happen to our results if the tuition of Geneva College was raised to 9.7 thousand dollars?" To see the effects of this change, we would open our Tuition workbook; select the Data worksheet tab; and select cell B4, the cell that contains the value for the tuition variable for Geneva; and enter the new value 9.7. Immediately, the value of the total cost variable changes from 13 to 13.8. Selecting the tab for the Results sheet, we notice no changes to the maximum, minimum, and range values. (You may wish to compare these results to the results produced by asking "What if the University of Pittsburgh's tuition, in cell B2, was changed from 10.3 to 13 thousand dollars?")

What if analyses are a valuable tool for understanding a variety of statistical methods. In Section 1S.15, we will learn about the scenario manager feature of Microsoft Excel, which facilitates the comparison of results among several, related what if analyses.

1S.15 USING THE SCENARIO MANAGER

When we wish to compare and save results produced by several related what ifs, or in cases in which we wish to preserve original values of our variables, we can use the **scenario manager** feature of Microsoft Excel. This feature allows us to associate a set of prestored values with a range of cells—called a scenario in Excel—and to save that set for later viewing or modification.

To illustrate the scenario manager, assume that we would like to determine and compare what happens to the value of the total cost range in each of the following four cases:

- Original data. Values as in Table 1S.1 on page 35.
- Pitt scholarship. We receive a scholarship that would pay for our tuition expenses for attending the University of Pittsburgh. (The tuition value for the University of Pittsburgh changes to zero.)
- ESU scholarship. We receive a scholarship that would pay for our tuition for attending East Stroudsburg University. (Tuition value for East Stroudsburg University changes to zero.)
- Drexel work/study. We receive a work/study grant that reduces the Drexel tuition to 3.0 thousand dollars. (The Drexel tuition value changes to 3.0.)

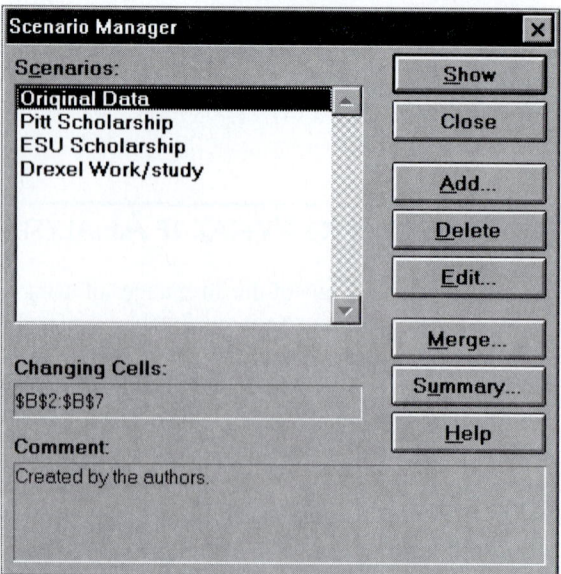

FIGURE IS.9 The Scenario Manager dialog box.

In this example we will need to define four scenarios, one for each of these cases, in order to compare results. To define the scenarios, we first select Tools | Scenarios, and the Scenario Manager dialog box appears (see Figure IS.9).

To define the four scenarios, we do the following:

❶ To define the original data scenario, click the Add button. The Add Scenario dialog box appears. Select the Scenario Name edit box and enter "Original Data" as the name for this scenario. Select the Changing Cells edit box and enter B2:B7. Click the OK button. The Scenario values dialog box appears, which contains edit boxes for each of the cells in the range. (Only the first five edit boxes are visible. Using the scroll bars would make the others visible.) As we want to store all the current values, immediately click the OK button. The scenario "original data" is now defined, and the Scenario Manager dialog box reappears.

❷ To define the Pitt scholarship scenario, click the Add button to get to the Add Scenario dialog box. Select the Scenario Name edit box and enter "Pitt Scholarship" as the name for this scenario. Verify that the Select the Changing Cells edit box contains B2:B7 (edit it, if necessary), and click the OK button. The Scenario Values dialog box appears. Locate the edit box for cell B2. Change the value in this box from 10.3 to 0 and then click the OK button. The scenario "Pitt scholarship" is now defined. Select Show the "Original Data" scenario to restore all the original data values.

❸ To define the ESU scholarship scenario, click the Add button to get to the Add Scenario dialog box. Select the Scenario Name edit box and enter "ESU Scholarship" as the name for this scenario. Verify that the Select the Changing Cells edit box contains B2:B7 (edit it, if necessary), and click the OK button. The Scenario Values dialog box appears. Locate the edit box for cell B3. Change the value in this box from 4.9 to 0 and then click the OK button. The scenario "ESU scholarship" is now defined. Select Show the "Original Data" scenario to restore all the original data values.

❹ To define the Drexel work/study scenario, again use a procedure similar to the previous two, naming the scenario "Drexel Work/study" and locating the edit box for cell

B5 in the Scenario Values dialog box. This time, as cell B7 is not one of the first five cells, we will have to scroll through the list of cells until it appears in the dialog box. Change the value in this box from 11.7 to 3 and then click the OK button. The scenario "Drexel work/study" is now defined. Select Show the "Original Data" scenario to restore all the original data values.

Having defined our scenarios, we can now use them in the following cycle:

① Select Tools | Scenarios if the Scenario Manager dialog box is not displayed.

② Select the scenario of interest and click the Show button, followed by the Close button.

③ Select the Results sheet tab to see the effects of the scenario.

④ Select Show the "Original Data" scenario to restore all the original data values.

⑤ Repeat steps 1–4 for each additional case.

All scenarios are saved when the workbook is saved, and they remain available until explicitly deleted (by pressing the Delete button in the Scenario Manager dialog box).

1S.16 IMPORTING DATA INTO A WORKSHEET

When performing statistical analyses throughout this text, we often encounter data sets with a large number of observations. When such sets have previously been entered and stored in a data file, it makes sense to try to import the contents of the file into our Data sheet to avoid having to reenter each observation one at a time as described in Section 1S.8.

Although Microsoft Excel can import data that have been stored in any one of various special file types used by other spreadsheet applications, in the field of statistics, data is more commonly found stored as **text files,** files that contain unlabeled and unformatted values that are separated by delimiters such as spaces, commas, or tab characters. When such a file is opened in Microsoft Excel, the program loads the **Text Import Wizard,** a series of linked dialog boxes that can be used to specify the organization of the data. As a user successfully steps through these dialog boxes, the Text Import Wizard will transfer the contents of the text file to a new worksheet in a new workbook. The new worksheet can then be transferred to another workbook using the File | Move or Copy Sheet command.

One text file found on the accompanying diskette is the TEXASC&U.TXT file, the out-of-state tuition data set of a population of 60 Texas colleges and universities that will be discussed extensively in Chapters 2 and 3. To import the data from this file, we would do the following:

TEXASC&U.TXT

① Select the command File | Open.

② Select the directory and drive that contains the text file TEXASC&U.TXT.

③ Enter the name TEXASC&U.TXT in the File Name edit box (or select it from the File Name list box) and click the OK button. The first dialog box of the Excel Text Wizard appears (see Figure 1S.10, Panel A).

④ Since the data values for the variables have been placed in aligned, fixed-width columns, select the Fixed Width option button. Press the Next button at the bottom of the dialog box to continue to the second dialog box of the wizard, illustrated in Figure 1S.10, Panel B. (Note that what the text calls variables are referred to as *fields* in the dialog box.)

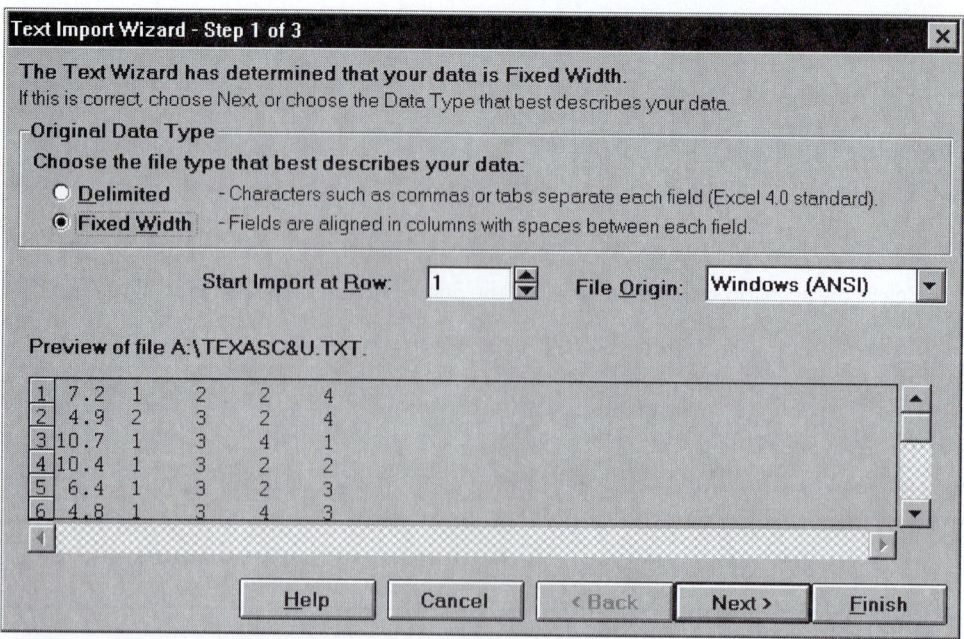

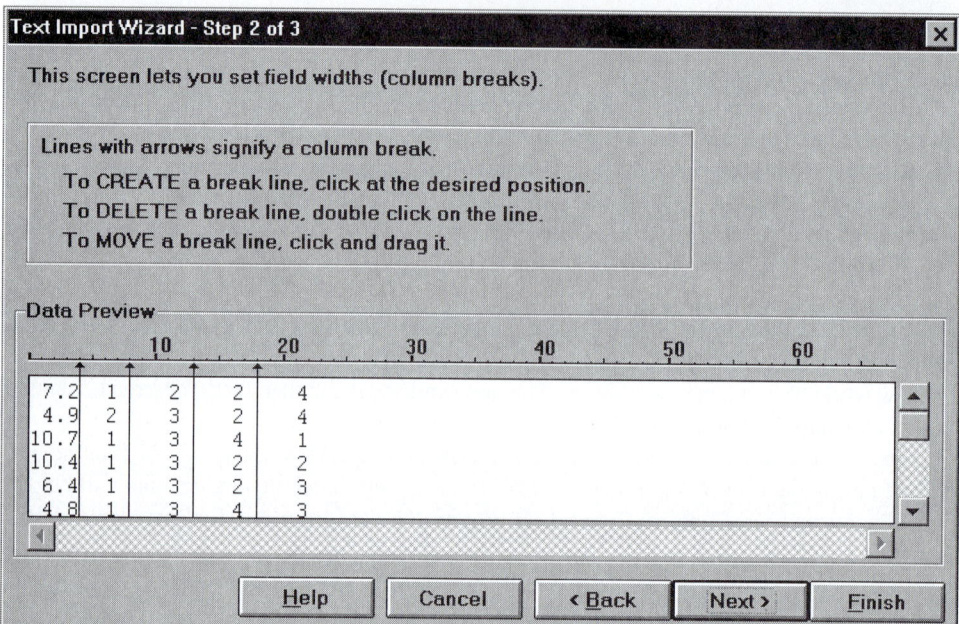

FIGURE IS.10 Using the Text Import Wizard for the Texas out-of-state tuition data. Panel A and Panel B.

⑤ This second dialog box (Panel B) allows you to specify how to place the data from each line in the text file into columns. In this case, accept the Text Import Wizard's suggested placement by pressing the Next button to continue to the third and last dialog box. (To override suggestions, click on the vertical arrow break line and drag it.)

⑥ This third dialog box (Panel C) allows you to select the format of the columns for each variable in the new worksheet and to designate which columns of data to exclude from the new worksheet, if any. In this case—as in most cases— accepting the

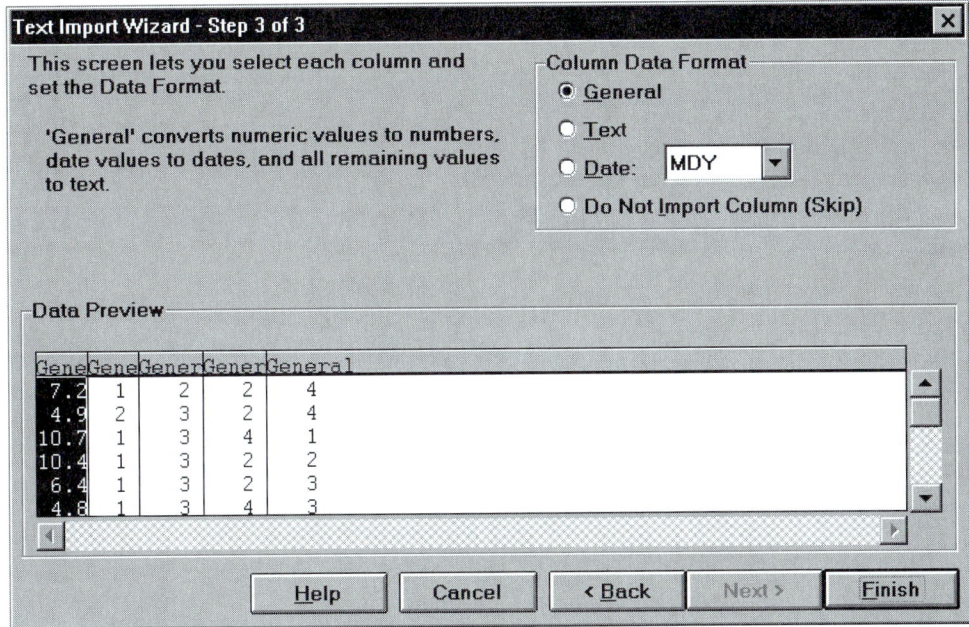

FIGURE 1S.10 Using the Text Import Wizard for the Texas out-of-state tuition data. Panel **C.**

Text Import Wizard's suggested "general" format for all columns, by clicking the Finish button, is all that needs to be done. (Otherwise, we would select a column and then select the option button corresponding to our choice for that column.)

Clicking the Finish button completes the text import process. At this point, you should review the new worksheet to check that the data have been correctly entered into all columns. After checking the data, select row 1 (by clicking on the row label 1 that is immediately to the left of cell A1) and select the command Insert | Rows. You will then have space to enter (and format) column headings as we did in Sections 1S.8 and 1S.11. Rename the worksheet "Data," save your workbook, and continue developing your workbook.

1S.17 SUMMARY

In this chapter we have provided an introduction to the basic concepts of spreadsheet applications in general and the features of the Microsoft Excel program in particular. We also used an example to illustrate the generation of a workbook-based solution to a simple statistical problem, by first analyzing the problem to be solved and then designing, implementing, and testing the solution.

Such a full discussion of all these steps for problems presented in later chapters is beyond the scope of this text. Instead, Microsoft Excel discussions in the remainder of this text will begin with detailed solution plans that need to be implemented. You should always remember, though, that thoughtful analysis preceded the development of those plans.

In the remainder of the text we will be learning about many additional aspects of Microsoft Excel in the context of specific statistical analyses. Our approach will be to integrate the discussion of Microsoft Excel with the statistical topics covered. In many instances, after covering a specific statistical procedure, we will demonstrate how to use Excel to analyze data that relates to the statistical methods just covered. In other situations, for pedagogical purposes, the discussion of Excel occurs after all the statistical topics in the chapter have been introduced.

Key Terms

Active cell 38	Operators 40
Bar 32	Option buttons 31
Buttons 32	Orientation 44
Cancel button 31	Overview sheet 34
Cell 37	Pull-down menu 32
Cell address 37	Renaming 35
Cell range 40	Right-clicking 29
Cell highlight 38	Scaling option buttons 44
Check box 31	Scenario manager 45
Clicking 29	Scroll bar 31
Data sheet 34	Scrollable list box 31
Dialog box 30	Select 29
Double-click 29	Sheet tab 35
Dragging 29	Spinner buttons 31
Drop-down list box 31	Spreadsheets 28
Edit box 30	Text files 47
Formula 39	Text Import Wizard 47
Function 42	Title bar 32
Icons 29	Toolbars 33
Menu bar 32	Windows 29
Mouse 29	Windowing environment 29
Mouse pointer 29	Workbook 28
OK button 31	Worksheet 28

References

1. Cobb Group, *Running Microsoft Excel 5* (Redmond, WA: Microsoft Press, 1994).

2. Grauer, R., and M. Barber, *Exploring Microsoft Excel 5.0* (Englewood Cliffs, NJ: Prentice-Hall, 1994).

3. *Microsoft Excel Version 7* (Redmond, WA: Microsoft Corp., 1996).

4. Parsons, J. J., Oja, D., and D. Auer, *Comprehensive Microsoft Excel 5.0 for Windows* (Cambridge, MA: Course Technology Inc., 1995).

5. Wells, E., *Developing Microsoft Excel 5 Solutions* (Redmond, WA: Microsoft Corp., 1995).

chapter 2

Presenting Data in Tables and Charts

To show how to organize and present collected data in tables and charts.

INTRODUCTION

In the preceding chapter we learned how to collect data through survey research. As pointed out in Section 1.9, since sampling saves time, money, and labor, we usually deal with sample information rather than data from an entire population. Nevertheless, regardless of whether we are dealing with a sample or a population, as a general rule, whenever a set of data that we have collected contains about 20 or more observations, the best way to examine such *mass data* is to present it in summary form by constructing appropriate tables and charts. We can then extract the important features of the data from these tables and charts.

Thus, this chapter is about data presentation. In particular, we will demonstrate how large sets of data can be organized and most effectively presented in the form of tables and charts in order to enhance data analysis and interpretation—two key aspects of the decision-making process.

2.2 ORGANIZING NUMERICAL DATA: THE ORDERED ARRAY AND STEM-AND-LEAF DISPLAY

In order to introduce the relevant ideas for Chapters 2 and 3, let us suppose that a company providing college advisory services to high school students throughout the United States has hired an analyst to compare the tuition rates charged to out-of-state residents by colleges and universities in different regions of the country. Table 2.1 displays the tuition rates charged to out-of-state residents by each of the 60 colleges and universities in the state of Texas (see Special Data Set 1 of Appendix D, pages D1–D2). When a set of data such as this one is collected, it is usually in **raw form**; that is, the numerical observations are not arranged in any particular order or sequence. As seen from Table 2.1, as the number of observations gets large, it becomes more and more difficult to focus on the major features in a set of data; thus, we need ways to organize the observations so that we can better understand what information the data are conveying. Two commonly used methods for accomplishing this are the *ordered array* and the *stem-and-leaf display*.

2.2.1 The Ordered Array

If we place the raw data in rank order, from the smallest to the largest observation, the ordered sequence obtained is called an **ordered array.** When the data are sorted into an ordered array,

Table 2.1   Raw data pertaining to tuition rates (in $000) for out-of-state residents at 60 colleges and universities in Texas.

TEXASC&U.TXT

7.2	4.9	10.7	10.4	6.4	4.8	4.7	4.6	6.0	5.4
4.8	4.7	8.3	3.8	4.8	8.3	6.4	6.6	4.5	8.0
3.6	2.4	8.5	8.8	7.7	4.9	8.6	12.0	4.9	7.0
11.0	4.9	3.9	4.9	4.4	4.9	4.9	8.0	3.6	7.4
7.9	4.9	5.8	3.9	11.6	10.3	3.4	3.9	5.0	3.9
8.0	3.5	4.9	5.8	4.1	3.9	3.5	4.8	5.9	3.6

Source: See Special Data Set 1, Appendix D, pages D1–D2, taken from "America's Best Colleges, 1994 College Guide," *U.S. News & World Report,* extracted from College Counsel 1993 of Natick, Mass. Reprinted by special permission, *U.S. News & World Report,* © 1993 by *U.S. News & World Report* and by College Counsel.

Table 2.2   Ordered array of out-of-state tuition rates (in $000) at 60 Texas colleges and universities.

2.4	3.4	3.5	3.5	3.6	3.6	3.6	3.8	3.9	3.9
3.9	3.9	3.9	4.1	4.4	4.5	4.6	4.7	4.7	4.8
4.8	4.8	4.8	4.9	4.9	4.9	4.9	4.9	4.9	4.9
4.9	4.9	5.0	5.4	5.8	5.8	5.9	6.0	6.4	6.4
6.6	7.0	7.2	7.4	7.7	7.9	8.0	8.0	8.0	8.3
8.3	8.5	8.6	8.8	10.3	10.4	10.7	11.0	11.6	12.0

Source: Table 2.1.

as in Table 2.2, our evaluation of their major features is facilitated. It becomes easier to pick out extremes, typical values, and concentrations of values.

Although it is useful to place the raw data into an ordered array prior to developing summary tables and charts or computing descriptive summary measures (see Chapter 3), the greater the number of observations present in a data set, the more useful it is to organize the data set into a stem-and-leaf display in order to study its characteristics (References 1, 13, and 14).

2.2.2   The Stem-and-Leaf Display

A **stem-and-leaf display** separates data entries into "leading digits" or "stems" and "trailing digits" or "leaves." For example, since the tuition rates (in $000) in the Texas data set all have one- or two-digit integer numbers, either the ones column or the tens column would be the leading digit, and the remaining column would be the trailing digit. Thus, an entry of 7.2 (corresponding to $7,200) has a leading digit of 7 and a trailing digit of 2.

Figure 2.1 depicts the stem-and-leaf display of the tuition rates for all 60 colleges and universities in Texas. The column of numbers to the left of the vertical line is called the "stem." These numbers correspond to the *leading digits* of the data. In each row the "leaves" branch out to the right of the vertical line, and these entries correspond to *trailing digits*.

● **Constructing the Stem-and-Leaf Display**   The stem-and-leaf display can be constructed using the data from Table 2.1. Note that the first institution, Abilene Christian University, has a tuition rate of 7.2 thousand dollars. Therefore, the trailing digit of 2 is listed as the first leaf value next to the stem value of 7 (the leading digit). The second institution, Angelo State University, has a tuition rate of 4.9 thousand dollars. Here the trailing digit of 9

```
 2 | 4
 3 | 869694995956
 4 | 9876878599994999918
 5 | 48089
 6 | 4046
 7 | 27049
 8 | 33058600
 9 |
10 | 743
11 | 06
12 | 0
```

$N = 60$

FIGURE 2.1   Stem-and-leaf display of out-of-state tuition rates at the Texas schools.
Source: Table 2.1.

is listed as the first leaf value next to the stem value of 4. Continuing, the third institution, Austin College, has a tuition rate of 10.7 thousand dollars so that the trailing digit of 7 is listed as the first leaf value next to the stem value of 10. The fourth institution, Baylor University, has a tuition rate of 10.4 thousand dollars, so the trailing digit of 4 is listed as the second leaf value next to the stem value of 10.

At this point in its construction, our stem-and-leaf display appears as follows:

```
 2 |
 3 |
 4 | 9
 5 |
 6 |
 7 | 2
 8 |
 9 |
10 | 74
11 |
12 |
```

Note that two of the four schools have the same stem. As more and more schools are included, those possessing the same stems and, perhaps, even the same leaves within stems (i.e., the same tuition rates) will be observed. Such leaf values will be recorded adjacent to the previously recorded leaves, opposite the appropriate stem—resulting in Figure 2.1.

To assist us in further examining the data, we may wish to rearrange the leaves within each of the stems by placing the digits in ascending order, row by row. The **revised stem-and-leaf display** is presented in Figure 2.2.

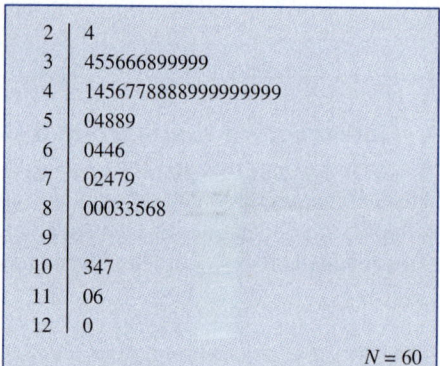

```
 2 | 4
 3 | 455666899999
 4 | 1456778888999999999
 5 | 04889
 6 | 0446
 7 | 02479
 8 | 00033568
 9 |
10 | 347
11 | 06
12 | 0
```

N = 60

FIGURE 2.2   **Revised stem-and-leaf display of out-of-state tuition rates at the Texas schools.**

Another type of rearrangement is also useful. If we desire to alter the size of the stem-and-leaf display, it is flexible enough for such an adjustment. Suppose, for example, we want to increase the number of stems so that we can attain a lighter concentration of leaves on the remaining stems. This is accomplished in the stem-and-leaf display presented in Figure 2.3.

Note that each stem from Figure 2.2 has been split into two new stems—one for the *low*-unit digits 0, 1, 2, 3, or 4 and the other for the *high*-unit digits 5, 6, 7, 8, or 9. These are represented by L and H, respectively, as indicated in the stem listings of Figure 2.3.

However, some researchers would argue that the data displayed in Figure 2.3 are undersummarized since we are failing to capture how the data are truly clustering within various

```
 2L │ 4
 2H │
 3L │ 4
 3H │ 55666899999
 4L │ 14
 4H │ 56778888999999999
 5L │ 04
 5H │ 889
 6L │ 044
 6H │ 6
 7L │ 024
 7H │ 79
 8L │ 00033
 8H │ 568
 9L │
 9H │
10L │ 34
10H │ 7
11L │ 0
11H │ 6
12L │ 0
                    N = 60
```

FIGURE 2.3 Revised stem-and-leaf display of out-of-state tuition rates at the Texas schools.
Source: Figure 2.2.

```
 2,3 │ 4455666899999
 4,5 │ 1456778888999999999904889
 6,7 │ 044602479
 8,9 │ 00033568
10,11 │ 34706
12,13 │ 0
                    N = 60
```

FIGURE 2.4 Revised stem-and-leaf display of out-of-state tuition rates at the Texas schools after condensing stems.
Source: Figure 2.2.

groupings. Hence, instead of expanding the display, as in Figure 2.3, we might wish to condense the data, as in Figure 2.4.

Note that consecutive pairs of stems from Figure 2.2 form the reduced set of stems in Figure 2.4 and the leaves corresponding to the *higher* member of each pair are **boldfaced.**

The (revised) stem-and-leaf display is, perhaps, the most versatile technique in descriptive statistics. It simultaneously organizes the data for further descriptive analyses (as we will see in Chapter 3), and it prepares the data for both tabular and chart form.

2.2.3 Using Microsoft Excel to Sort Data into an Ordered Array

In this section we began to study the out-of-state tuition rates of 60 Texas colleges by forming an ordered array and a stem-and-leaf display. Although the stem-and-leaf display is not available in Microsoft Excel,[1] we can use the Data | Sort command to obtain an ordered array. To obtain an ordered array similar to Table 2.2 on page 53, begin by opening the TEXAS-1.XLS workbook. Click on the Data worksheet tab. You will observe that the Texas data consists of six variables that have been labeled School, Tuition, Type, Setting, Calendar, and Focus. Since we want to sort by the tuition variable, click anywhere in Column B and select the command Data | Sort.

TEXAS-1.XLS

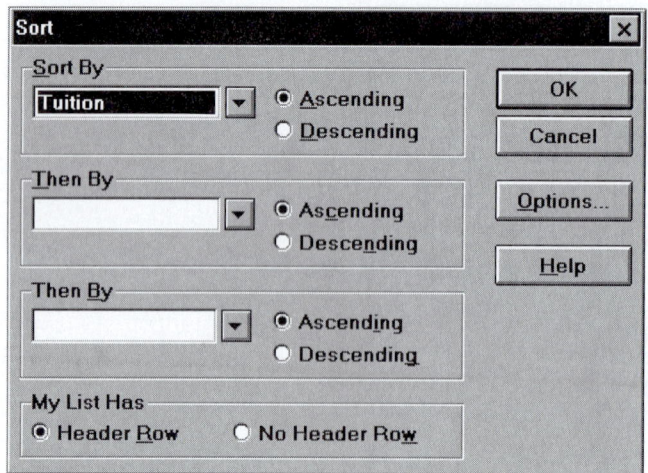

FIGURE 2.1.EXCEL Sort dialog box for the Texas out-of-state tuition data.

The Sort dialog box presented in Figure 2.1.Excel appears. In the upper left portion of this dialog box, Excel provides a list box that lists the names of all the variables in your data. These names were obtained from the column headings for each column. Excel allows you to sort the data according to one, two, or three variables. The first variable that the data are to be sorted by is entered at the top followed by (if desired) the second and third variables by which the data should be sorted. For each variable to be sorted, you have the choice of sorting in ascending order (lowest to highest) or descending order (highest to lowest). Since we want to obtain an ordered array for the tuition variable, verify that tuition appears in the Sort By box and that the first ascending option button has been selected. You also will notice that the Header Row button has been selected in this dialog box. This will allow the data to be sorted without moving the column headings. Click the OK button to sort the data. Figure 2.2.Excel represents the Texas data sorted in ascending order by the amount of tuition.

Problems for Section 2.2

● 2.1 Given the following stem-and-leaf display:

9	714
10	82230
11	561776735
12	394282
13	20

(a) Rearrange the leaves and form the revised stem-and-leaf display.
(b) Place the data into an ordered array.
(c) Which of these two devices seems to give more information? Discuss.

MAILORD.TXT

2.2 Upon examining the monthly billing records of a mail-order book company, the auditor takes a sample of 20 of its unpaid accounts. The amounts owed the company are

$4, $18, $11, $7, $7, $10, $5, $33, $9, $12

$3, $11, $10, $6, $26, $37, $15, $18, $10, $21

(a) Develop the ordered array.
(b) Form the stem-and-leaf display.

	A	B	C	D	E	F
1	School	Tuition	Type	Setting	Calendar	Focus
2	Prairie View A&M U.	2.4	Public	Rural	Semester	RU
3	U. of Houston	3.4	Public	Urban	Semester	NU
4	U. of Texas at Arlington	3.5	Public	Suburban	Semester	NU
5	U. of Texas, San Antonio	3.5	Public	Urban	Semester	RU
6	Paul Quinn C.	3.6	Private	Urban	Semester	RLA
7	Texas C.	3.6	Private	Urban	Semester	RLA
8	Wiley C.	3.6	Private	Urban	Semester	RLA
9	Jarvis Christian C.	3.8	Private	Rural	Semester	RLA
10	Sul Ross State U.	3.9	Public	Rural	Semester	RU
11	Texas Women's U.	3.9	Public	Urban	Semester	RU
12	U. of Houston-Downtown	3.9	Public	Urban	Semester	RU
13	U. of North Texas	3.9	Public	Urban	Semester	RU
14	U. of Texas-Pan American	3.9	Public	Urban	Semester	RU
15	U. of Texas at El Paso	4.1	Public	Urban	Semester	RU
16	Texas A&I U.	4.4	Public	Rural	Semester	RU
17	Midwestern State U.	4.5	Public	Urban	Semester	RU
18	East Texas State U.	4.6	Private	Urban	Quarter	RU
19	East Texas Baptist U.	4.7	Private	Urban	4-1-4	RLA
20	Huston-Tillotson C.	4.7	Private	Urban	Semester	RLA
21	Dallas Baptist U.	4.8	Private	Urban	4-1-4	RLA
22	Howard Payne U.	4.8	Private	Rural	Semester	RLA
23	Lamar U.	4.8	Public	Urban	Semester	RU
24	Wayland Baptist U.	4.8	Private	Urban	4-1-4	RU
25	Angelo State U.	4.9	Public	Urban	Semester	RU
26	Sam Houston State U.	4.9	Public	Rural	Semester	RU
27	Southwest Texas State U.	4.9	Public	Urban	Semester	RU
28	Stephen F. Austin State U.	4.9	Public	Rural	Semester	RU
29	Tarleton State U.	4.9	Public	Rural	Semester	RU
30	Texas A&M U.	4.9	Public	Urban	Semester	NU
31	Texas A&M at Galveston	4.9	Public	Urban	Semester	RU
32	Texas Tech U.	4.9	Public	Urban	Semester	RU
33	U. of Texas at Austin	4.9	Public	Urban	Semester	NU
34	U. of Mary Hardin-Baylor	5	Private	Suburban	Semester	RLA
35	Houston Baptist U.	5.4	Private	Urban	Quarter	RU
36	Texas Wesleyan U.	5.8	Private	Urban	Semester	RU
37	U. of Texas at Dallas	5.8	Public	Suburban	Semester	RU
38	West Texas State U.	5.9	Public	Rural	Semester	RU
39	Hardin Simmons U.	6	Private	Urban	Semester	RU
40	Concordia Lutheran C.	6.4	Private	Urban	Semester	RLA
41	Lubbock Christian U.	6.4	Private	Urban	Semester	RLA
42	McMurry U.	6.6	Private	Urban	4-1-4	RLA
43	Southwestern Adventist	7	Private	Rural	Semester	RLA
44	Abilene Christian U.	7.2	Private	Suburban	Semester	RU
45	Texas Lutheran C.	7.4	Private	Urban	Semester	RLA
46	St. Mary's U.	7.7	Private	Urban	Calendar	RU
47	Texas Southern U.	7.9	Public	Urban	Semester	RU
48	Our Lady of the Lake	8	Private	Urban	Semester	RU
49	Texas Christian U.	8	Private	Urban	Semester	NU
50	U. of St. Thomas	8	Private	Urban	Semester	RU
51	Incarnate Word C.	8.3	Private	Urban	Semester	RLA
52	LeTourneau U.	8.3	Private	Urban	Semester	RLA
53	Rice U.	8.5	Private	Urban	Semester	NU
54	Schreiner C.	8.6	Private	Rural	4-1-4	RLA
55	St. Edward's U.	8.8	Private	Urban	Semester	RU
56	U. of Dallas	10.3	Private	Suburban	Semester	NLA
57	Baylor U.	10.4	Private	Urban	Semester	NU
58	Austin C.	10.7	Private	Urban	4-1-4	NLA
59	Southwestern U.	11	Private	Suburban	Semester	RLA
60	Trinity U.	11.6	Private	Urban	Semester	RU
61	Southern Methodist U.	12	Private	Suburban	Semester	NU

FIGURE 2.2.EXCEL Sorted Texas out-of-state tuition data.

SCALES.TXT

2.3 The following data represent the retail price (in dollars) of a sample of 39 different brands of bathroom scales:

50	50	50	28	65	40	50	22	32	30
79	50	22	20	35	24	25	120	35	35
65	20	14	25	24	48	15	10	17	50
25	22	60	30	12	30	10	12	20	

Source: "Bathroom Scales," Copyright 1993 by Consumers Union of United States, Inc., Yonkers, N.Y. 10703. Adapted by permission from *Consumer Reports*, January 1993, pp. 34–35. Although these data sets originally appeared in *Consumer Reports*, the selective adaptation and resulting conclusions presented are those of the authors and are not sanctioned or endorsed in any way by Consumers Union, the publisher of *Consumer Reports*.

(a) Develop the ordered array.
(b) Form the stem-and-leaf display.

STOCK1.TXT

● 2.4 The following data are the book values (in dollars) (i.e., net worth divided by number of outstanding shares) for a random sample of 50 stocks from the New York Stock Exchange (NYSE):

7	9	8	6	12	6	9	15	9	16
8	5	14	8	7	6	10	8	11	4
10	6	16	5	10	12	7	10	15	7
10	8	8	10	18	8	10	11	7	10
7	8	15	23	13	9	8	9	9	13

(a) Develop the ordered array.
(b) Form the stem-and-leaf display.

CANCER.TXT

2.5 A medical doctor speaking on a late-night television show conjectures that "cancer appears to be more prevalent in states with large urban populations and in states in the eastern part of the United States." The following data represent the incidence rate of cancer (i.e., reported incidence per 100,000 population) in all 50 states during a recent year:

State	Incidence of Cancer per 100,000 Population	State	Incidence of Cancer per 100,000 Population
Alabama	433	Kentucky	414
Alaska	442	Louisiana	422
Arizona	360	Maine	391
Arkansas	383	Maryland	491
California	366	Massachusetts	443
Colorado	282	Michigan	454
Connecticut	434	Minnesota	366
Delaware	500	Mississippi	438
Florida	367	Missouri	390
Georgia	406	Montana	372
Hawaii	371	Nebraska	336
Idaho	307	Nevada	422
Illinois	402	New Hampshire	403
Indiana	438	New Jersey	464
Iowa	377	New Mexico	375
Kansas	345	New York	329

(continued)

(continued)

State	Incidence of Cancer per 100,000 Population	State	Incidence of Cancer per 100,000 Population
North Carolina	355	Tennessee	408
North Dakota	408	Texas	313
Ohio	463	Utah	229
Oklahoma	326	Vermont	376
Oregon	396	Virginia	440
Pennsylvania	442	Washington	364
Rhode Island	445	West Virginia	409
South Carolina	418	Wisconsin	398
South Dakota	348	Wyoming	238

Source: National Cancer Institute.

(a) Develop the ordered array.
(b) Form the stem-and-leaf display.

2.6 The following data represent the type (creamy versus chunky), score (0 = poor, 100 = excellent), cost (in cents), and amount of sodium (in mgs) of a sample of 37 brands of peanut butter:

PEANUT.XLS

Product	Type	Score	Cost (¢)	Sodium (mgs)
Jif	Creamy	68	22	220
Smucker's Natural	Creamy	65	27	15
Deaf Smith Arrowhead Mills	Creamy	62	32	0
Adams 100% Natural	Creamy	56	26	0
Adams	Creamy	56	26	168
Skippy	Creamy	56	19	225
Laura Scudder's All Natural	Creamy	53	26	165
Kroger	Creamy	50	14	240
Country Pure Brand (Safeway)	Creamy	50	21	225
NuMade (Safeway)	Creamy	45	20	187
Peter Pan	Creamy	44	21	225
Peter Pan (2)	Creamy	41	22	3
A&P	Creamy	40	12	225
Hollywood Natural	Creamy	40	32	15
Food Club	Creamy	39	17	225
Pathmark	Creamy	36	9	255
Lady Lee (Lucky Stores)	Creamy	30	16	225
Albertsons	Creamy	30	17	225
Shur Fine (Shurfine Central Corp.)	Creamy	22	16	225
Smucker's Natural	Chunky	80	27	15
Jif	Chunky	75	23	162
Skippy	Chunky	75	21	211
Adams 100% Natural	Chunky	62	26	0
Deaf Smith Arrowhead Mills	Chunky	62	32	0
Country Pure Brand (Safeway)	Chunky	62	21	195
Laura Scudder's All Natural	Chunky	56	24	165
Smucker's Natural	Chunky	53	26	188

(continued)

(continued)

Product	Type	Score	Cost (¢)	Sodium (mgs)
Food Club	Chunky	52	17	195
Kroger	Chunky	50	14	255
A&P	Chunky	47	11	225
Peter Pan	Chunky	47	22	180
NuMade (Safeway)	Chunky	42	21	208
Health Valley 100% Natural	Chunky	42	34	3
Lady Lee (Lucky Stores)	Chunky	40	16	225
Albertsons	Chunky	36	17	225
Pathmark	Chunky	34	9	210
Shur Fine (Shurfine Central Corp.)	Chunky	34	16	195

Source: "Peanut Butter," Copyright 1990 by Consumers Union of United States, Inc., Yonkers, N.Y. 10703. Adapted by permission of *Consumer Reports,* September 1990, p. 590. Although these data sets originally appeared in *Consumer Reports*, the selective adaptation and resulting conclusions presented are those of the authors and are not sanctioned or endorsed in any way by Consumers Union, the publisher of *Consumer Reports*.

For each of the three variables (score, cost, and sodium)
(a) Use the Sort command in Microsoft Excel to develop the ordered array.
(b) Form the stem-and-leaf display.

SHAMPOO.TXT

2.7 The following data are the cost per ounce (in cents) for random samples of 31 conventional shampoos labeled for "normal" hair and 29 shampoos labeled for "fine" hair:

Normal Hair					Fine Hair				
79	63	19	9	37	69	9	23	22	8
49	20	16	55	69	12	32	12	18	74
23	14	9	87	44	19	63	49	37	55
13	16	23	20	64	85	44	87	17	11
28	18	32	81	85	23	50	65	51	35
47	50	8	13	21	14	20	28	8	
9									

Source: "Shampoos," Copyright 1992 by Consumers Union of United States, Inc., Yonkers, N.Y. 10703. Adapted by permission from *Consumer Reports,* June 1992, pp. 400–401. Although these data sets originally appeared in *Consumer Reports*, the selective adaptation and resulting conclusions presented are those of the authors and are not sanctioned or endorsed in any way by Consumers Union, the publisher of *Consumer Reports*.

(a) Develop the ordered array for each data set.
(b) Form the stem-and-leaf display for each data set.

2.3 TABULATING NUMERICAL DATA: THE FREQUENCY DISTRIBUTION

Using either the raw data, an ordered array, or a revised stem-and-leaf display of out-of-state tuition rates for the 60 colleges and universities in Texas (see Tables 2.1 and 2.2 on pages 52 and 53 and Figure 2.1 on page 53), the analyst wishes to construct the appropriate tables and charts that will enhance the report she is preparing for the marketing manager of the college advisory service.

Regardless of whether an ordered array or a stem-and-leaf display is selected for *organizing* the data, as the number of observations gets large, it becomes necessary to further condense the data into appropriate summary tables. Thus, we may wish to arrange the data into **class groupings** (i.e., categories) according to conveniently established divisions of the range of the observations. Such an arrangement of data in tabular form is called a frequency distribution.

> A **frequency distribution** is a summary table in which the data are arranged into conveniently established, numerically ordered class groupings or categories.

When the observations are *grouped* or condensed into frequency distribution tables, the process of data analysis and interpretation is made much more manageable and meaningful. In such summary form the major data characteristics can be approximated, thus compensating for the fact that when the data are so grouped, the initial information pertaining to individual observations that was previously available is lost through the grouping or condensing process.

In constructing the frequency distribution table, attention must be given to

1. Selecting the appropriate *number* of class groupings for the table
2. Obtaining a suitable *class interval* or *width* of each class grouping
3. Establishing the *boundaries* of each class grouping to avoid overlapping

2.3.1 Selecting the Number of Classes

The number of class groupings to be used is primarily dependent on the number of observations in the data. Larger numbers of observations require a larger number of class groups. In general, however, the frequency distribution should have at least five class groupings, but no more than 15. If there are not enough class groupings or if there are too many, little information would be obtained. As an example, a frequency distribution having but one class grouping that spans the entire range of tuition rates could be formed as follows:

Tuition Rates (in $000)	Number of Schools
2.0–13.0	60
Total	60

From such a summary table, however, no additional information is obtained that was not already known from scanning either the raw data or the ordered array. A table with too much data concentration is not meaningful. The same would be true at the other extreme—if a table had too many class groupings, there would be an underconcentration of data, and very little would be learned.

2.3.2 Obtaining the Class Intervals

When developing the frequency distribution table, it is desirable to have each class grouping of equal width. To determine the width of each class, the *range* of the data is divided by the number of class groupings desired:

$$\text{Width of interval} \cong \frac{\text{range}}{\text{number of desired class groupings}} \qquad (2.1)$$

Since there are only 60 observations in our tuition rate data, we decided that six class groupings would be sufficient. From the ordered array in Table 2.2 (page 53), the range is computed as 12.0 – 2.4 = 9.6 thousand dollars, and, using Equation (2.1), the **width of the class interval** is approximated by

$$\text{Width of interval} \cong \frac{9.6}{6} = 1.6 \text{ thousand dollars}$$

For convenience and ease of reading, the selected interval or width of each class grouping is rounded to 2.0 thousand dollars.

2.3.3 Establishing the Boundaries of the Classes

To construct the frequency distribution table, it is necessary to establish clearly defined **class boundaries** for each class grouping so that the observations either in raw form or in an ordered array can be properly tallied. Overlapping of classes must be avoided.

Since the width of each class interval for the tuition rate data has been set at 2.0 thousand dollars, the boundaries of the various class groupings must be established so as to include the entire range of observations. Whenever possible, these boundaries should be chosen to facilitate the reading and interpreting of data. Thus, the first class interval ranges from 2.0 to under 4.0, the second from 4.0 to under 6.0, and so on. The data in their raw form (Table 2.1) or from the ordered array (Table 2.2) are then tallied into each class as shown:

Tuition Rates (in $000)	Tallies	Frequency
2.0 but less than 4.0	~~HH~~ ~~HH~~ ///	13
4.0 but less than 6.0	~~HH~~ ~~HH~~ ~~HH~~ ~~HH~~ ////	24
6.0 but less than 8.0	~~HH~~ ////	9
8.0 but less than 10.0	~~HH~~ ///	8
10.0 but less than 12.0	~~HH~~	5
12.0 but less than 14.0	/	1
Total		60

By establishing the boundaries of each class as above, all 60 observations have been tallied into six classes, each having an interval width of 2.0 thousand dollars, without overlapping. The frequency distribution is presented in Table 2.3.

Table 2.3 **Frequency distribution of the Texas out-of-state tuition rates.**

Tuition Rates (in $000)	Number of Schools
2.0 but less than 4.0	13
4.0 but less than 6.0	24
6.0 but less than 8.0	9
8.0 but less than 10.0	8
10.0 but less than 12.0	5
12.0 but less than 14.0	1
Total	60

Source: Table 2.1.

The main advantage of using this summary table is that the major data characteristics become immediately clear to the reader. For example, we see from Table 2.3 that the *approximate range* of the 60 tuition rates is from 2.0 to 14.0 thousand dollars, with out-of-state tuition at most Texas schools tending to cluster between 4.0 and 6.0 thousand dollars.

On the other hand, the major disadvantage of this summary table is that we cannot know how the individual values are distributed within a particular class interval without access to the original data. Thus, for the five schools with out-of-state tuition rates between 10.0 and 12.0 thousand dollars, it is not clear from Table 2.3 whether the values are distributed throughout the interval, cluster near 10.0 thousand dollars, or cluster near 12.0 thousand dollars. The class midpoint, however, is the value used to represent all the data summarized into a particular interval.

The **class midpoint** is the point halfway between the boundaries of each class and is representative of the data within that class.

The class midpoint for the interval "2.0 but less than 4.0" is 3.0 thousand dollars. (The other class midpoints are, respectively, 5.0, 7.0, 9.0, 11.0, and 13.0 thousand dollars.)

2.3.4   Subjectivity in Selecting Class Boundaries

The selection of class boundaries for frequency distribution tables is highly subjective. Hence, for data sets that do not contain many observations, the choice of a particular set of class boundaries over another might yield a different picture to the reader. For example, for the tuition rate data, using a class-interval width of 2.5 thousand dollars instead of 2.0 (as was used in Table 2.3) may cause shifts in the way in which the observations distribute among the classes. This is particularly true if the number of observations in the data set is not very large.

Such shifts in data concentration do not occur only because the width of the class interval is altered. We may keep the interval width at 2.0 thousand dollars but choose different lower and upper class boundaries. Such manipulation may also cause shifts in the way in which the data distribute—especially if the size of the data set is not very large. Fortunately, as the number of observations in a data set increases, alterations in the selection of class boundaries affect the concentration of data less and less.

Problems for Section 2.3

Note: *To use Microsoft Excel to solve these problems, refer to Section 2.7.*

2.8   A random sample of 50 executive vice-presidents is selected from various public relations firms in the United States, and the annual salaries of these company officers are obtained. The salaries range from $52,000 to $137,000. Set up the class boundaries for a frequency distribution
(a) If 5 class intervals are desired
(b) If 6 class intervals are desired
(c) If 7 class intervals are desired
(d) If 8 class intervals are desired

2.9   If the asking price of one-bedroom cooperative and condominium apartments in Queens, a borough of New York City, varies from $103,000 to $295,000,
(a) Indicate the class boundaries of 10 classes into which these values can be grouped.
(b) What class-interval width did you choose?
(c) What are the 10 class midpoints?

● 2.10   The raw data displayed here are the electric and gas utility charges during the month of July 1995 for a random sample of 50 three-bedroom apartments in Manhattan:

UTILITY.TXT

Raw Data on Utility Charges ($)

96	171	202	178	147	102	153	197	127	82
157	185	90	116	172	111	148	213	130	165
141	149	206	175	123	128	144	168	109	167
95	163	150	154	130	143	187	166	139	149
108	119	183	151	114	135	191	137	129	158

(a) Form a frequency distribution
    (1) Having 5 class intervals
    (2) Having 6 class intervals
    (3) Having 7 class intervals
    [*Hint:* To help you decide how best to set up the class boundaries, you should first place the raw data either in a stem-and-leaf display (by letting the leaves be the trailing digits) or in an ordered array.]
(b) Form a frequency distribution having 7 class intervals with the following class boundaries: $80 but less than $100, $100 but less than $120, and so on.

2.11 Construct a frequency distribution from the book value data in Problem 2.4 on page 58.

2.12 Construct a frequency distribution from the cancer incidence data in Problem 2.5 on pages 58–59.

2.13 Construct separate frequency distributions for each of the three numerical variables (score, cost, and sodium) from the peanut butter data in Problem 2.6 on pages 59–60.

2.14 Given the ordered arrays in the accompanying table dealing with the lengths of life (in hours) of a sample of forty 100-watt light bulbs produced by Manufacturer A and a sample of forty 100-watt light bulbs produced by Manufacturer B:

BULBS.TXT

Ordered arrays of length of life of two brands of 100-watt light bulbs (in hours).

Manufacturer A

684	697	720	773	821
831	835	848	852	852
859	860	868	870	876
893	899	905	909	911
922	924	926	926	938
939	943	946	954	971
972	977	984	1005	1014
1016	1041	1052	1080	1093

Manufacturer B

819	836	888	897	903
907	912	918	942	943
952	959	962	986	992
994	1004	1005	1007	1015
1016	1018	1020	1022	1034
1038	1072	1077	1077	1082
1096	1100	1113	1113	1116
1153	1154	1174	1188	1230

(a) Form the frequency distribution for each brand. (*Hint:* For purposes of comparison, choose class-interval widths of 100 hours for each distribution.)

(b) For purposes of answering Problems 2.19, 2.24, and 2.30, form the frequency distribution for each brand according to the following schema [if you have not already done so in part (a) of this problem]:

Manufacturer A: 650 but less than 750, 750 but less than 850, and so on

Manufacturer B: 750 but less than 850, 850 but less than 950, and so on

(c) Change the class-interval width in (b) to 50 so that you have intervals from 650 to under 700, 700 to under 750, 750 to under 800, and so on.

2.4 TABULATING NUMERICAL DATA: THE RELATIVE FREQUENCY DISTRIBUTION AND PERCENTAGE DISTRIBUTION

The frequency distribution is a summary table into which the original data are grouped to facilitate data analysis. To enhance the analysis, however, it is almost always desirable to form either the relative frequency distribution or the percentage distribution, depending on whether we prefer proportions or percentages. These two equivalent distributions are shown as Tables 2.4 and 2.5, respectively.

The **relative frequency distribution** depicted in Table 2.4 is formed by dividing the frequencies in each class of the frequency distribution (Table 2.3 on page 62) by the total number of observations. A **percentage distribution** (Table 2.5) may then be formed by multiplying each relative frequency or proportion by 100.0. Thus, from Table 2.4 it is clear that the proportion of schools in Texas with out-of-state tuition rates from 12.0 to under 14.0 thousand dollars is .017, while from Table 2.5 it can be seen that 1.7% of the schools have such tuition rates.

Working with a base of 1 for proportions or 100.0 for percentages is usually more meaningful than using the frequencies themselves. Indeed, the use of the relative frequency distribution or percentage distribution becomes essential whenever one set of data is being compared with other sets of data, especially if the numbers of observations in each set differ.

As a case in point, let us suppose that a personnel manager wanted to compare daily absenteeism among the clerical workers in two department stores. If, on a given day, 6 clerical workers out of 50 in Store A were absent and 3 clerical workers out of 10 in Store B were absent, what conclusions can be drawn? It is inappropriate to say that *more* absenteeism occurred in Store A. Although there were twice as many absences in Store A as there were in

Table 2.4 Relative frequency distribution of the Texas out-of-state tuition rates.

Tuition Rates (in $000)	Proportion of Schools
2.0 but less than 4.0	.217
4.0 but less than 6.0	.400
6.0 but less than 8.0	.150
8.0 but less than 10.0	.133
10.0 but less than 12.0	.083
12.0 but less than 14.0	.017
Total	1.000

Source: Data are taken from Table 2.3 on page 62.

Table 2.5 Percentage distribution of the Texas out-of-state tuition rates.

Tuition Rates (in $000)	Percentage of Schools
2.0 but less than 4.0	21.7
4.0 but less than 6.0	40.0
6.0 but less than 8.0	15.0
8.0 but less than 10.0	13.3
10.0 but less than 12.0	8.3
12.0 but less than 14.0	1.7
Total	100.0

Source: Data are taken from Table 2.3 on page 62.

Store B, there were also five times as many clerical workers employed in Store A. Hence, in these types of comparisons, we must formulate our conclusions from the *relative rates* of absenteeism, not from the actual counts. Thus, we can state that the absenteeism rate is two and a half times higher in Store B (30.0%) than it is in Store A (12.0%).

Now suppose that, when developing her report for the marketing manager of the college advisory service, our analyst decides to compare the 60 out-of-state Texas tuition rates with those reported from 45 higher education institutions in the state of North Carolina. Table 2.6 displays information on the out-of-state resident tuition rate for *each* of the 45 North Carolina colleges and universities (see Special Data Set 1 of Appendix D on page D3).

To compare the out-of-state tuition rates from the 60 Texas schools with those from the 45 North Carolina schools, we develop a percentage distribution for the latter group. This new table will then be compared with Table 2.5.

Table 2.7 depicts both the frequency distribution and the percentage distribution of the tuition rates charged to out-of-state residents by the 45 North Carolina schools. This table has been constructed instead of two separate tables to save space. Note that the class groupings selected in Table 2.7 match, where possible, those selected in Table 2.3 for the Texas schools. The boundaries of the classes should match or be multiples of each other in order to facilitate comparisons.

Using the percentage distributions of Tables 2.5 and 2.7, it is now meaningful to compare the schools in the two states in terms of the tuition rates charged to out-of-state residents. From the

NCC&U-T.TXT

Table 2.6 Raw data pertaining to tuition rates (in $000) for out-of-state residents at 45 colleges and universities in North Carolina.

6.5	4.0	7.1	8.3	5.4	7.6	9.0	15.7	16.7
6.4	5.0	8.5	5.7	7.7	7.2	12.4	7.1	5.5
9.7	4.4	7.0	6.3	8.3	6.9	5.7	7.6	7.9
7.9	6.0	8.2	10.4	9.9	3.9	9.8	8.2	5.6
7.9	6.4	7.4	7.0	13.0	8.7	6.4	6.7	7.4

Source: See Special Data Set 1, Appendix D, page D3, taken from "America's Best Colleges, 1994 College Guide," *U.S. News & World Report*, extracted from College Counsel 1993 of Natick, Mass. Reprinted by special permission, *U.S. News & World Report*, © 1993 by *U.S. News & World Report* and by College Counsel.

Table 2.7 Frequency distribution and percentage distribution of North Carolina out-of-state tuition rates.

Tuition Rates (in $000)	Number of Schools	Percentage of Schools
2.0 but less than 4.0	1	2.2
4.0 but less than 6.0	8	17.8
6.0 but less than 8.0	21	46.7
8.0 but less than 10.0	10	22.2
10.0 but less than 12.0	1	2.2
12.0 but less than 14.0	2	4.4
14.0 but less than 16.0	1	2.2
16.0 but less than 18.0	1	2.2
Totals	45	99.9*

*Error due to rounding.
Source: Data are taken from Table 2.6.

two tables it is apparent that the tuition rates are generally lower in Texas than in North Carolina. For example, in Texas, tuition rates are typically clustering between 4.0 and 6.0 thousand dollars (i.e., 40.0% of the schools) while in North Carolina the tuition rates are typically clustering between 6.0 and 8.0 thousand dollars (or 46.7% of the schools). We can also approximate the *ranges* in tuition rates from the tables. In North Carolina, the range in tuition rates is approximated to be 16.0 thousand dollars (i.e., the difference between 18.0, the upper boundary of the last class, and 2.0, the lower boundary of the first class), while in Texas the range is approximated as 12.0 thousand dollars (or 14.0 – 2.0). Other descriptive summary measures that would enhance a comparative analysis of the tuition rates between the two states will be discussed in Chapter 3.

Problems for Section 2.4

Note: *To use Microsoft Excel to solve these problems, refer to Section 2.7.*

- 2.15 Form the percentage distribution from the frequency distribution developed in Problem 2.10(b) on page 64 regarding utility charges.
- 2.16 Form the percentage distribution from the frequency distribution developed in Problem 2.11 on page 64 regarding book values of companies listed on the NYSE.
  2.17 Form the percentage distribution from the frequency distribution developed in Problem 2.12 on page 64 regarding cancer incidence.
  2.18 Form the percentage distributions corresponding to the frequency distributions for each of the three numerical variables (score, cost, and sodium) developed in Problem 2.13 on page 64 regarding peanut butter characteristics.

S-2-18.XLS

  2.19 Form the percentage distributions from the frequency distributions developed in Problem 2.14(b) on page 65 concerning the life of light bulbs manufactured by two competing companies, A and B.

2.5 GRAPHING NUMERICAL DATA: THE HISTOGRAM AND POLYGON

It is often said that "one picture is worth a thousand words." Indeed, statisticians often employ graphic techniques to more vividly describe sets of data. In particular, histograms and polygons are used to describe numerical data that have been grouped into frequency, relative frequency, or percentage distributions.

2.5.1 Histograms

Histograms are vertical bar charts in which the rectangular bars are constructed at the boundaries of each class.

When plotting histograms, the random variable or phenomenon of interest is displayed along the horizontal axis; the vertical axis represents the number, proportion, or percentage of observations per class interval—depending on whether the particular histogram is, respectively, a frequency histogram, a relative frequency histogram, or a percentage histogram.

Vertical Axis Label	↔	Type of Chart
Number of observations	↔	Frequency histogram (or polygon)
Proportion of observations	↔	Relative frequency histogram (or polygon)
Percentage of observations	↔	Percentage histogram (or polygon)

A percentage histogram is depicted in Figure 2.5 for the out-of-state tuition rates at all 60 colleges and universities in Texas. It is interesting to note the close visual relationship portrayed by the stem-and-leaf display and the histogram. Look at Figure 2.4 on page 55 and our histogram in Figure 2.5. If we were to rotate the stem-and-leaf display 90° (i.e., hold our book sideways), a frequency histogram would be depicted so that its class groupings would be represented by the stems and its vertical bars would be represented by individual leaves on each stem.

When comparing two or more sets of data, neither stem-and-leaf displays nor histograms can be constructed on the same graph. Superimposing the vertical bars of one histogram on another would cause difficulty in interpretation. For such cases, it is necessary to construct relative frequency or percentage polygons.

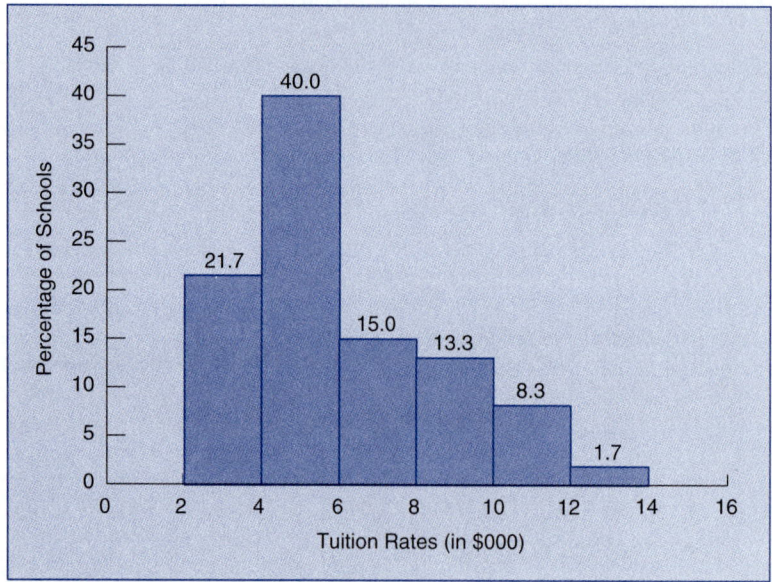

FIGURE 2.5 **Percentage histogram of the Texas out-of-state tuition rates.**
Source: Data are taken from Table 2.5 on page 65.

2.5.2 Polygons

As with histograms, when plotting polygons the phenomenon of interest is displayed along the horizontal axis, and the vertical axis represents the number, proportion, or percentage of observations per class interval.

The **percentage polygon** is formed by letting the midpoint of each class represent the data in that class and then connecting the sequence of midpoints at their respective class percentages.

Because consecutive midpoints are connected by a series of straight lines, the polygon is sometimes jagged in appearance. However, when dealing with a very large set of data, if we were to make the boundaries of the classes in its frequency distribution closer together (and thereby increase the number of classes in that distribution), the jagged lines of the polygon would "smooth out."

Figure 2.6 shows the percentage polygon for the out-of-state tuition rates at all 60 Texas schools, and Figure 2.7 compares the percentage polygons for the tuition rates at the 60 Texas schools versus the 45 North Carolina schools. The differences in the structure of the two distributions, previously discussed when comparing Tables 2.5 and 2.7, are clearly indicated here.

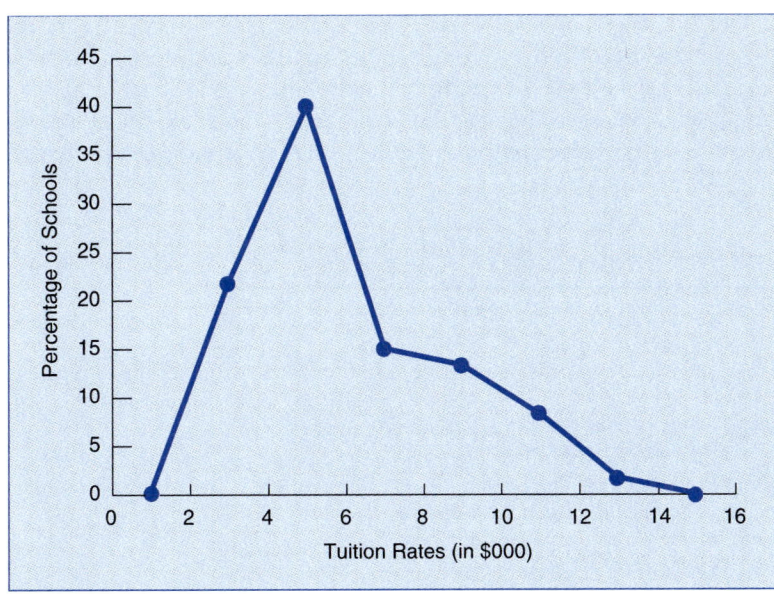

FIGURE 2.6   **Percentage polygon of the Texas out-of-state tuition rates.**
Source: Data are taken from Table 2.5.

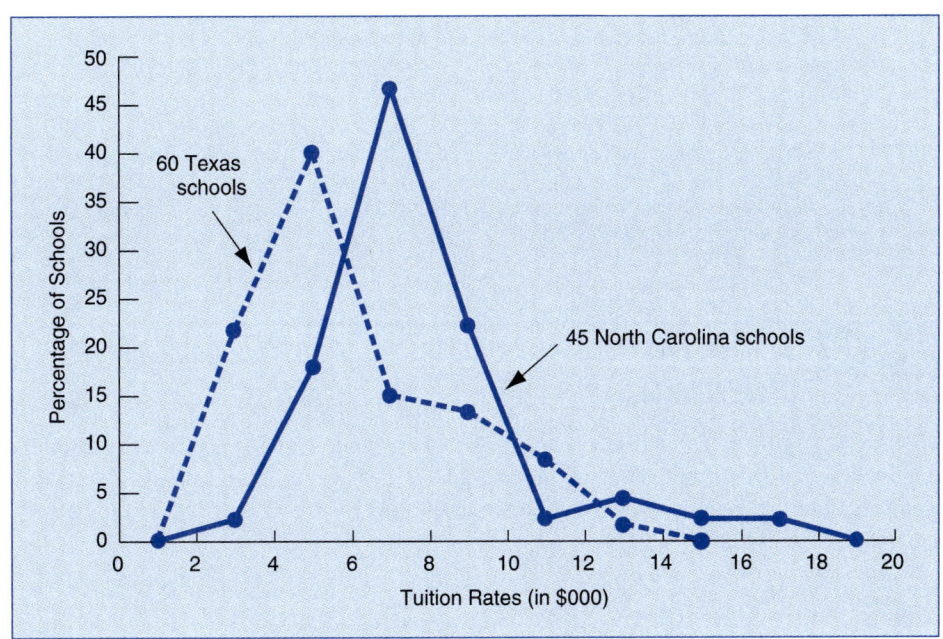

FIGURE 2.7   **Percentage polygons of out-of-state tuition rates for Texas and North Carolina schools.**   *Source:* Data are taken from Tables 2.5 and 2.7.

● **Polygon Construction**   Notice that the polygon is a representation of the shape of the particular distribution. Since the area under the percentage distribution (the entire curve) must be 100.0%, it is necessary to connect the first and last midpoints with the horizontal axis so as to enclose the area of the observed distribution. In Figure 2.6, this is accomplished by connecting the first observed midpoint with the midpoint of a "fictitious preceding" class (i.e., 1.0 thousand dollars) having 0.0% observations and by connecting the last observed midpoint with the midpoint of a "fictitious succeeding" class (i.e., 15.0 thousand dollars) having 0.0% observations.

Notice, too, that when polygons (Figure 2.6) or histograms (Figure 2.5) are constructed, the vertical axis must show the true zero or "origin" so as not to distort or otherwise misrepresent the character of the data. The horizontal axis, however, does not need to specify the zero point for the phenomenon of interest. For aesthetic reasons, the range of the random variable should constitute the major portion of the chart, and, when zero is not included, "breaks" (─╲╱─) in the axis are appropriate.

Problems for Section 2.5

Note: *To use Microsoft Excel to solve these problems, refer to Section 2.7.*

UTILITY.TXT

2.20 From the percentage distribution developed in Problem 2.15 on page 67 regarding utility charges
(a) Plot the percentage histogram.
(b) Plot the percentage polygon.

STOCK1.TXT

2.21 From the percentage distribution developed in Problem 2.16 on page 67 regarding book values of companies listed on the NYSE
(a) Plot the percentage histogram.
(b) Plot the percentage polygon.

CANCER.TXT

2.22 From the percentage distribution developed in Problem 2.17 on page 67 regarding cancer incidence
(a) Plot the percentage histogram.
(b) Plot the percentage polygon.

S-2-23.XLS

2.23 From the percentage distributions developed in Problem 2.18 on page 67 for each of the three numerical variables (score, cost, and sodium) regarding peanut butter characteristics
(a) Plot the respective percentage histograms.
(b) Plot the respective percentage polygons.

BULBS.TXT

2.24 From the percentage distributions developed in Problem 2.19 on page 67 regarding life of light bulbs
(a) Plot the percentage histograms on separate graphs.
(b) Plot the percentage polygons on one graph.

2.6 CUMULATIVE DISTRIBUTIONS AND CUMULATIVE POLYGONS

Two other useful methods of data presentation that facilitate analysis and interpretation are the cumulative distribution tables and the cumulative polygon charts. Both of these may be developed from the frequency distribution table, the relative frequency distribution table, or the percentage distribution table.

2.6.1 The Cumulative Percentage Distribution

Depending on our individual preference for proportions or percentages, when comparing two or more sets of data of differing size, we select either the relative frequency distribution or the percentage distribution. Since we already have the percentage distributions of out-of-state tuition rates at 60 Texas schools and at 45 North Carolina schools in Tables 2.5 and 2.7 (pages 65 and 66), we can use these tables to construct the respective cumulative percentage distributions. See Tables 2.8 and 2.9.

A **cumulative percentage distribution table** is constructed by first recording the lower boundaries of each class from the percentage distribution and then inserting an extra boundary at the end. We compute the cumulative percentages in the *"less than"* column by determining the percentage of observations less than each of the stated boundary values. Thus from

Table 2.8 Cumulative percentage distribution of out-of-state tuition rates for the 60 Texas schools.

Tuition Rates (in $000)	Percentage of Schools "Less Than" Indicated Value
2.0	0.0
4.0	21.7
6.0	61.7
8.0	76.7
10.0	90.0
12.0	98.3
14.0	100.0

Source: Data are taken from Table 2.5.

Table 2.9 Cumulative percentage distribution of out-of-state tuition rates for the 45 North Carolina schools.

Tuition Rates (in $000)	Percentage of Schools "Less Than" Indicated Value
2.0	0.0
4.0	2.2
6.0	20.0
8.0	66.7
10.0	88.9
12.0	91.1
14.0	95.6
16.0	97.8
18.0	100.0

Source: Data are taken from Table 2.7.

Table 2.5, we see that 0.0% of the out-of-state tuition rates in Texas institutions are less than 2.0 thousand dollars; 21.7% of the tuition rates are less than 4.0 thousand dollars; 61.7% of the tuition rates are less than 6.0 thousand dollars; and so on until all (100.0%) of the tuition rates are less than 14.0 thousand dollars. This cumulating process is observed in Table 2.10.

Table 2.10 Forming the cumulative percentage distribution.

Out-of-State Tuition Rates (in $000)	Percentage of Schools in Class Interval	Percentage of Schools "Less Than" Lower Boundary of Class Interval
2.0 but less than 4.0	21.7	0.0
4.0 but less than 6.0	40.0	21.7
6.0 but less than 8.0	15.0	61.7 = 21.7 + 40.0
8.0 but less than 10.0	13.3	76.7 = 21.7 + 40.0 + 15.0
10.0 but less than 12.0	8.3	90.0 = 21.7 + 40.0 + 15.0 + 13.3
12.0 but less than 14.0	1.7	98.3 = 21.7 + 40.0 + 15.0 + 13.3 + 8.3
14.0 but less than 16.0	0.0	100.0 = 21.7 + 40.0 + 15.0 + 13.3 + 8.3 + 1.7

2.6.2 Cumulative Percentage Polygon

To construct a **cumulative percentage polygon** (also known as an **ogive**), we note that the phenomenon of interest—tuition rates—is again plotted on the horizontal axis, while the cumulative percentages (from the "*less than*" column) are plotted on the vertical axis. At each lower boundary, we plot the corresponding (cumulative) percentage value from the listing in the cumulative percentage distribution. We then connect these points with a series of straight-line segments.

Figure 2.8 on page 72 illustrates the cumulative percentage polygon of the Texas out-of-state tuition rates. The major advantage of the ogive over other charts is the ease with which we can interpolate between the plotted points.

● **Approximating the Percentages**   As one example, the analyst at the college advisory service might wish to approximate the percentage of Texas colleges and universities

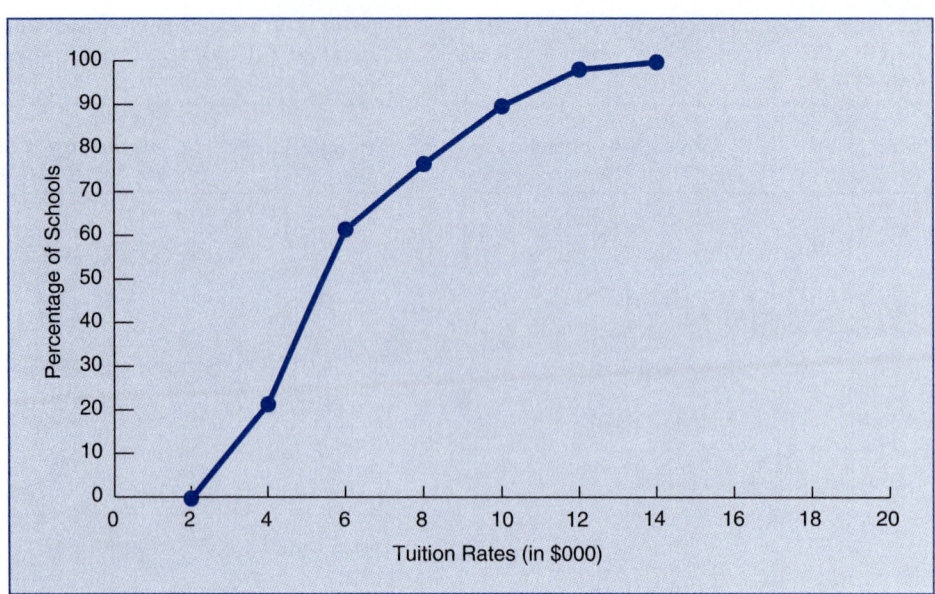

FIGURE 2.8 Cumulative percentage polygon of the Texas out-of-state tuition rates.
Source: Data are taken from Table 2.8.

that charge an out-of-state tuition rate below a specified amount, say 7.0 thousand dollars. To accomplish this, a vertical line is projected upward at 7.0 until it intersects the "less than" curve. The desired percentage is then approximated by reading horizontally from the point of intersection to the percentage indicated on the vertical axis. In this case, approximately 69.2% of the Texas schools have tuition rates under 7.0 thousand dollars. (This, of course, implies that about 30.8% of the schools have tuition rates of at least 7.0 thousand dollars.)

● **Approximating the Values** Even more important, the analyst, when preparing her report for the marketing manager of the college advisory service, may also wish to approximate various tuition rates that correspond to particular cumulative percentages. For example, 25.0% of all Texas schools have out-of-state tuition rates below what amount? To determine this, a horizontal line is drawn from the specified cumulative percentage point (25.0) until it intersects the "less than" curve. The desired tuition rate is then approximated by dropping a perpendicular (a vertical line) at the point of intersection to the horizontal axis. From Figure 2.8, we note that this rate is approximately 4.2 thousand dollars. Other percentage points commonly considered for such analysis (see Chapter 3) are the 50.0% value and the 75.0% value.

● **Comparing Two or More Cumulative Distributions** Such approximations as these are extremely helpful when comparing two or more sets of data. Figure 2.9 depicts the cumulative percentage polygons of out-of-state tuition rates for both the Texas schools and the North Carolina schools.

From Figure 2.9, we note that, in general, the Texas ogive is drawn to the left of the North Carolina ogive. For example, in Texas, 25% of all tuition rates are below 4.2 thousand dollars, while in North Carolina, we see that 25% of all tuition rates are below 6.1 thousand dollars. In Texas, 50% of all tuition rates are below 5.4 thousand dollars, while in North Carolina, 50% of all tuition rates are below 7.2 thousand dollars. Furthermore, in Texas, 75% of all tuition rates are below 7.7 thousand dollars, while in North Carolina, we see that 75% of all tuition rates are below 8.7 thousand dollars. These comparisons enable us to confirm our earlier impression that out-of-state tuition rates are lower in Texas than in North Carolina.

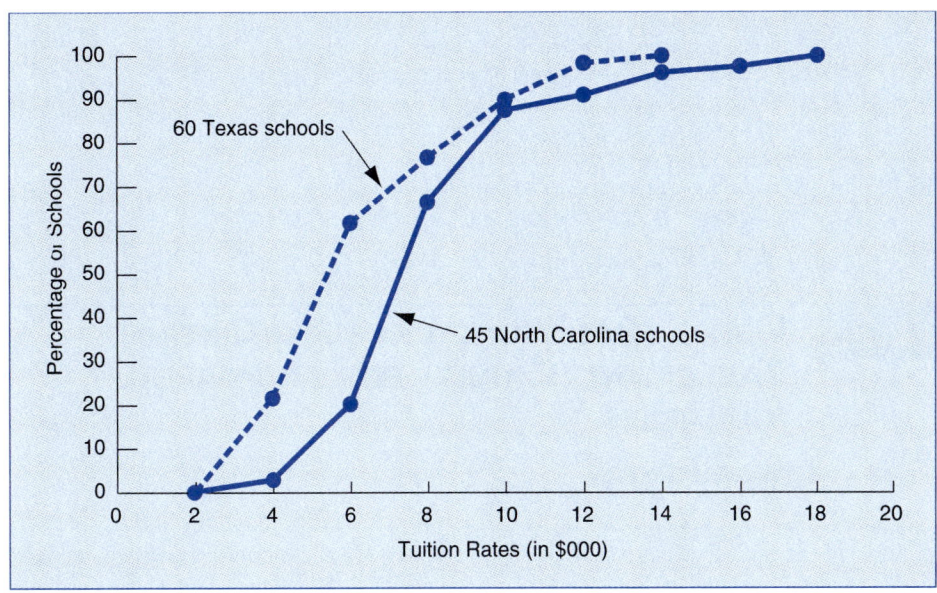

FIGURE 2.9 **Cumulative percentage polygons of out-of-state tuition rates for the Texas and North Carolina schools.** *Source:* Data are taken from Tables 2.5 and 2.7.

Problems for Section 2.6

Note: *To use Microsoft Excel to solve these problems, refer to Section 2.7.*

2.25 Examine Figure 2.9.

(a) 10.0% of the out-of-state tuition rates in each state are below what amounts?

(b) 40.0% of the out-of-state tuition rates in each state are below what amounts?

(c) 60.0% of the out-of-state tuition rates in each state are below what amounts?

(d) 90.0% of the out-of-state tuition rates in each state are below what amounts?

(e) What percentage of the out-of-state tuition rates in each state are below 5.0 thousand dollars?

(f) What percentage of the out-of-state tuition rates in each state are below 11.0 thousand dollars?

(g) Discuss your findings.

(h) How might your information be of assistance to the research analyst at the college advisory service? Discuss.

TEXAS-2.XLS
NCC&U.TXT

● 2.26 From the frequency distribution developed in Problem 2.10(b) on page 64 regarding utility charges

(a) Form the cumulative frequency distribution.

(b) Form the cumulative percentage distribution.

(c) Plot the ogive (cumulative percentage polygon).

UTILITY.TXT

● 2.27 From the frequency distribution developed in Problem 2.11 on page 64 regarding book values of companies listed on the NYSE

(a) Form the cumulative frequency distribution.

(b) Form the cumulative percentage distribution.

(c) Plot the ogive (cumulative percentage polygon).

STOCK1.TXT

2.28 From the frequency distribution developed in Problem 2.12 on page 64 regarding cancer incidence

(a) Form the cumulative frequency distribution.

(b) Form the cumulative percentage distribution.

(c) Plot the ogive (cumulative percentage polygon).

CANCER.TXT

2.29 From the frequency distributions developed in Problem 2.13 on page 64 for each of the three numerical variables (score, cost, and sodium) regarding peanut butter characteristics

S-2-29.XLS

(a) Form the respective cumulative frequency distributions.
(b) Form the respective cumulative percentage distributions.
(c) Plot the respective ogives (cumulative percentage polygons).

BULBS.TXT

2.30 From the frequency distributions developed in Problem 2.14 on pages 64–65 regarding life of light bulbs from two manufacturers
(a) Form the cumulative frequency distributions.
(b) Form the cumulative percentage distributions.
(c) Plot the ogives (cumulative percentage polygons) on one graph.

2.7 USING MICROSOFT EXCEL TO OBTAIN TABLES AND CHARTS FOR NUMERICAL VARIABLES

2.7.1 Introduction

In Sections 2.3–2.6, we organized the Texas out-of-state tuition data into a frequency distribution and then formed relative frequency, percentage, and cumulative percentage distributions. Using these distributions, we then developed the histogram, percentage polygon, and cumulative percentage polygon. In this section we will discuss how Microsoft Excel can be used to obtain tables and charts for these data. To organize our workbook for this purpose, we need to design a Data sheet and a Calculations sheet. Additional sheets including those for the histogram and frequency polygon charts will be developed as a consequence of selecting the appropriate Excel command. Table 2.1.Excel represents the design of the Data sheet, while Table 2.2.Excel represents the design of the Calculations sheet.

TEXAS-1.XLS

2.7.2 Using the FREQUENCY Function to Obtain a Frequency Distribution

One way to obtain a frequency distribution in Excel is to use the special array formulas that contain the FREQUENCY function of Excel. We can develop formulas in the form:

$$\text{=FREQUENCY } (cell\ range,\ upper\ class\ boundaries)$$

Table 2.1.Excel Design of the Data sheet for the Texas out-of-state tuition workbook.

	A	B	C	D	E	F	G
1	School	Tuition	Type	Setting	Calendar	Focus	Upper Class
2	xx	xx	xx	xx	xx	xx	Boundaries
3	.	.	.	.	.	.	1.99
4	.	.	.	.	.	.	3.99
5	.	.	.	.	.	.	5.99
6	.	.	.	.	.	.	7.99
7	.	.	.	.	.	.	9.99
8	.	.	.	.	.	.	11.99
9	.	.	.	.	.	.	13.99
⋮	.	.	.	.	.	.	
61	xx	xx	xx	xx	xx	xx	

Table 2.2.Excel Design of the Calculations sheet for the Texas out-of-state tuition workbook.

	A	B	C	D
1		Texas Frequency Distribution		
2				
3	Class	Frequency	Relative Frequency	Percentage
4	=DATA!G3	{=FREQUENCY(DATA!B2:B61,DATA!G3:G9)}	=B4/$B$11	=C4
5	=DATA!G4	.	=B5/$B$11	=C5
6	=DATA!G5	.	=B6/$B$11	=C6
7	=DATA!G6	.	=B7/$B$11	=C7
8	=DATA!G7	.	=B8/$B$11	=C8
9	=DATA!G8	.	=B9/$B$11	=C9
10	=DATA!G9	.	=B10/$B$11	=C10
11	Total:	=SUM(B4:B10)		

where   *cell range* = the cell range for the data to be analyzed

upper class boundaries = the cell range that contains the values that represent the upper class boundaries

To create a frequency distribution for the Texas tuition data similar to Table 2.3 on page 62, open your Texas tuition workbook or the TEXAS-1.XLS file and do the following:

❶ Select the Data sheet and enter the upper class boundaries in the cell range G3:G9 that corresponds to the class intervals shown in Table 2.3. For this problem, they are the values 1.99, 3.99, 5.99, 7.99, 9.99, 11.99, and 13.99. We have written 1.99 as the approximation of less than 2, 3.99 as the approximation of less than 4, ..., and 13.99 as the approximation of less than 14. These limits need to be entered on the Data sheet when we use the Data Analysis tool in Section 2.7.3.

❷ Issue the command Insert | Worksheet and name the worksheet as Calculations.

❸ Enter title and column headings for this sheet.

❹ In the cell range A4:A10, enter the formulas to copy the values of the upper class boundaries. As before, you can enter =Data!G3 into cell A4 and copy the formula to the remainder of the range (see Section 1S.10).

❺ We are now ready to assemble the formula to calculate the frequencies. As the values for the Texas tuition data are located in cells B2:B61 on the Data sheet and the upper class boundaries are located in the range G3:G9 on the same sheet, we can use the formula =FREQUENCY(Data!B2:B61,Data!G3:G9). To enter this formula, we take advantage of a shortcut feature of Excel that will place this formula into a range of cells in one operation. To do this, first select the range for the formula (in this case, B4:B10 of the Calculations sheet). Then type the formula and, *while holding down the Control and Shift keys*, press the Enter key. The frequencies now appear on Column B on this sheet (see Figure 2.3.Excel). If you use the formula bar to review the formulas just entered in the range B4:B10, you will notice that the formulas are enclosed in curly braces {}. This indicates that these cells contain a special type of formula that cannot be individually edited.

	A	B
	B4	{=F
1		Texas Frequenc
2		
3	Class	Frequency
4	1.99	0
5	3.99	13
6	5.99	24
7	7.99	9
8	9.99	8
9	11.99	5
10	13.99	1

FIGURE 2.3.EXCEL Frequency distribution for the Texas out-of-state tuition data obtained from the Excel FREQUENCY function.

▲ WHAT IF EXAMPLE

We can store alternative sets of upper class boundaries and compare the results using the Scenario Manager feature (see Section 1S.15). Define scenarios for the range Data!G3:G9 using the upper class boundaries of this chapter (1.99, 3.99, 5.99, 7.99, 9.99, 11.99, 13.99) as well as the set of values of .99, 2.99, 4.99, 6.99, 8.99, 10.99, and 12.99. Additional scenarios in which the class-interval width is some other value than two are also possible. However, these scenarios would require the editing of the Data and Calculations sheets. A frequency distribution with this different class-interval width would be obtained and could be compared to the one illustrated in Figure 2.3.Excel.

Having obtained the set of frequencies, we can now calculate the relative frequencies and percentages. As a first step, enter the formula =SUM(B4:B10) in cell B11 of the Calculations sheet to obtain the total frequency. (The formula is entered into cell B11 for presentation purposes.) This value can now be used as the denominator in a set of formulas in column C to calculate the relative frequencies. To enter the relative frequencies for the first class, enter the formula =B4/$B$11 in cell C4 and press Enter. In this formula, the address in the denominator has been entered as an **absolute address**. This is an address that will *not* be adjusted by Excel during the copying operation. (This makes sense here because we want to divide the cell frequencies by the same value, the total frequency, which is contained in cell B11.)

The relative frequencies are now displayed in cells C4:C10. They can be converted to percentages by doing the following:

❶ Copy cell C4 to cell D4 by entering the formula =C4 in cell D4.

❷ Copy this formula through the range D4:D10.

❸ Select the cell range D4:D10 and press the Percent format button.

❹ Adjust the decimal by pressing the Increase Decimal button once for each decimal place desired. In Figure 2.4.Excel, we pressed the Increase Decimal button twice to obtain two decimal places.

The frequencies, relative frequencies, and percentages are illustrated in Figure 2.4.Excel.

	A	B	C	D
1	Texas Frequency Distribution			
2				
3	Class	Frequency	Relative Freq.	Percentage
4	1.99	0	0	0.00%
5	3.99	13	0.216666667	21.67%
6	5.99	24	0.4	40.00%
7	7.99	9	0.15	15.00%
8	9.99	8	0.133333333	13.33%
9	11.99	5	0.083333333	8.33%
10	13.99	1	0.016666667	1.67%
11	Total:	60		

FIGURE 2.4.EXCEL Frequency distribution, relative frequency distribution, and percentage distribution obtained from Excel for the Texas out-of-state tuition data.

2.7.3 Using the Data Analysis Tool to Obtain Frequency and Cumulative Frequency Distributions and Histograms

In addition to (or instead of) using the FREQUENCY function to obtain frequency distributions, we could use the Histogram option of the Data Analysis tool of Excel to obtain frequency and cumulative frequency distributions as well as a histogram and cumulative frequency polygon. The Data Analysis tool is a set of predefined routines that we can use to perform most of the statistical procedures we will discuss in this text. These routines allow us to obtain results that are often difficult or impossible to obtain using simple formulas. The Data Analysis tool also typically generates numerous statistics from a single-user operation. However, the statistics generated by this tool do not change when the data being analyzed changes. In such circumstances, it may be better to use formulas where possible. For this reason, we discuss both the use of the Data Analysis tool and the use of formulas.

The Histogram option of the Data Analysis tool can be used to obtain both a frequency distribution and charts such as a histogram and cumulative percentage polygon. To use the Histogram option of the Data Analysis tool, the upper class boundaries of the class intervals *must* be entered on the sheet that contains the data to be analyzed. Retrieve the TEXAS-2.XLS workbook file, which is the original Texas out-of-state tuition data as shown in Table 2.1.Excel, and select the Data Sheet by clicking on the Data Sheet tab. To create the histogram and cumulative percentage polygon, we need to do the following:

TEXAS-2.XLS

❶ Select the command Tools | Data Analysis and then select Histogram from the Analysis Tools list box that appears (see Figure 2.5.Excel). Click the OK button to display the Histogram dialog box. If Data Analysis is not a choice on your Tools menu, the Data Analysis component of Excel is probably not properly installed (review Section 1S.4 before continuing).

❷ In the Input area of the Histogram dialog box, enter the range Data!$B$2:$B$61 in the Input Range edit box. Enter the range Data!$G$3:$G$9 in the Bin Range edit box.

❸ In the Output area of the Histogram dialog box, click the New Worksheet Ply button and enter Histogram as the name of the new sheet in the edit box to the right of this button. Select the Cumulative Percentages and Chart Output check boxes and leave

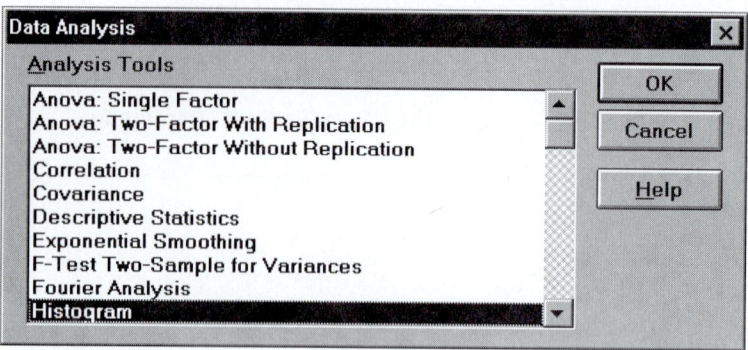

FIGURE 2.5.EXCEL   Data Analysis dialog box.

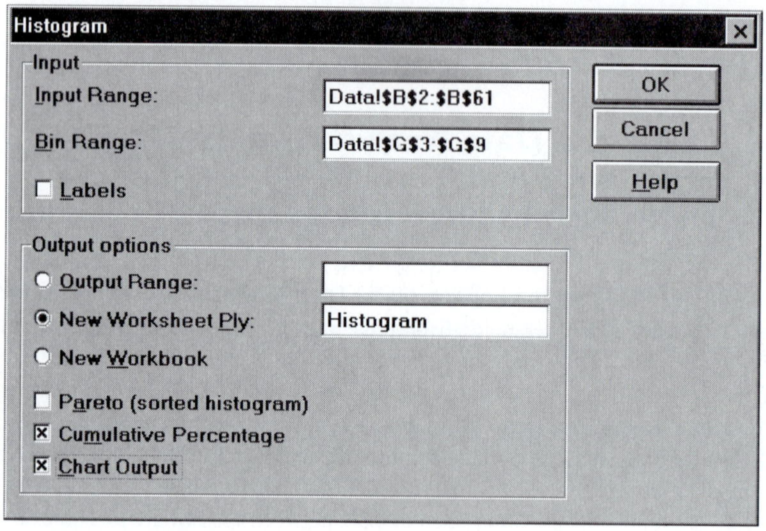

FIGURE 2.6.EXCEL   Histogram dialog box. Panel A.

the Pareto check box unselected. Click the OK button. The dialog box should look like the one shown in Figure 2.6.Excel, Panel A.

Excel will generate both a frequency distribution and cumulative percentage distribution and superimpose the cumulative percentage polygon onto the histogram (see Figure 2.6.Excel, Panel B). If only a histogram was desired, the Cumulative Percentages check box would not be selected in step 3.

Observe that as was the case with the FREQUENCY function, the frequencies and cumulative percentages provided refer to the upper boundaries of the class. This means that 21.67% of the schools have tuition less than 4 thousand dollars, 61.67% have tuition less than 6 thousand dollars, 76.67% have tuition less than 8 thousand dollars, 90% have tuition less than 10 thousand dollars, 98.33% have tuition less than 12 thousand dollars, and 100% have tuition less than 14 thousand dollars.

In Figure 2.6.Excel Panel B, a different vertical axis is included for each chart since the two graphs are superimposed. The vertical axis on the left side of the chart provides frequencies for the histogram, while the vertical axis on the right provides percentages for the cumulative percentage polygon.

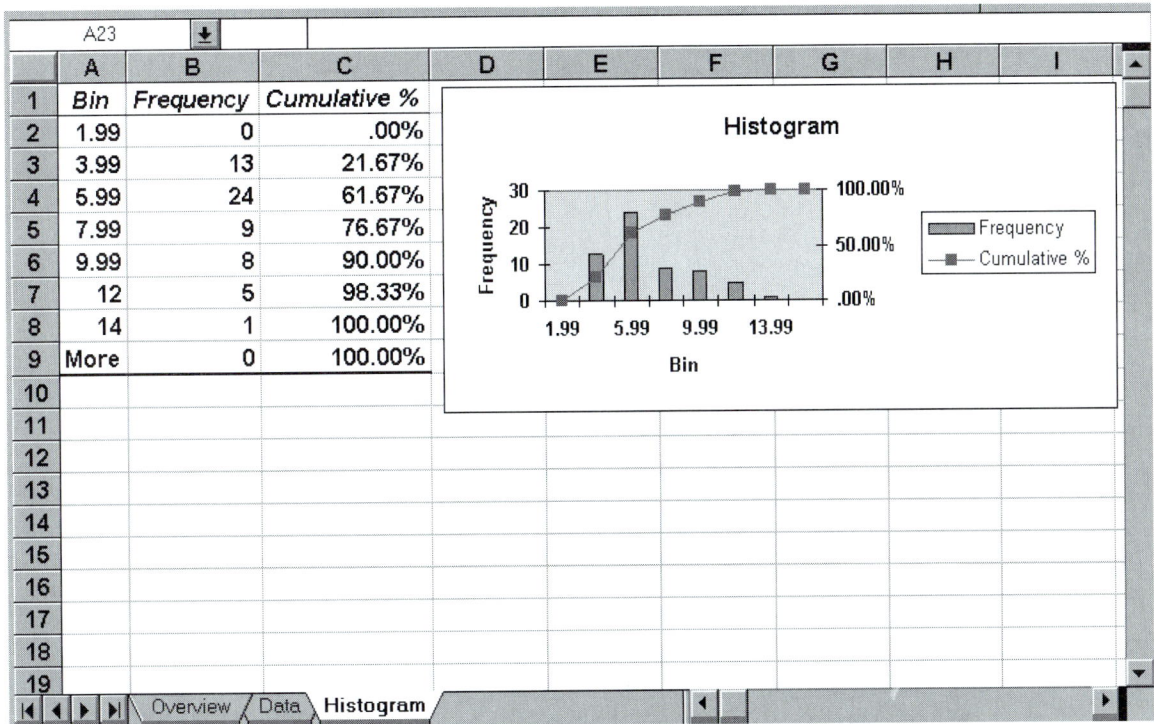

	A	B	C	D	E	F	G	H	I
1	Bin	Frequency	Cumulative %						
2	1.99	0	.00%						
3	3.99	13	21.67%						
4	5.99	24	61.67%						
5	7.99	9	76.67%						
6	9.99	8	90.00%						
7	12	5	98.33%						
8	14	1	100.00%						
9	More	0	100.00%						

FIGURE 2.6.EXCEL Histogram and cumulative percentage distribution obtained from Excel for the Texas out-of-state tuition data. Panel B.

Observe that this chart contains two errors. There are gaps between the bars that correspond to the class intervals, and there is an additional class, labeled More by Excel. To remove the gaps, we need to do the following:

❶ Double-click on the white area in the chart box.

❷ Select Format | 1 Column Group. This displays the Format Column Groups dialog box.

❸ In the Gap Width edit box of the Options tab, change the value to 0. Click the OK button. The Histogram now has continuous bars.

To remove the additional class, we need to do the following:

❶ First select a cell outside the chart. Then double-click on one of the bars displayed on the histogram. A halo appears around the chart box, and a second set of data points appears superimposed over the bars along with a formula that begins with the word Series in the edit box above the worksheet.

❷ Select the command Format | Selected Data Series to display the Format Data Series dialog box. Click the Name and Values tab. In the Y Values edit box, change the ending cell from $B$9 to $B$8. Click the OK button.

❸ Then click on one of the data points of the Cumulative Percentage chart. Select Format | Selected Data Series a second time, and the Name and Values tab appears; change the ending cell in the Y Values edit box, from $C$9 to $C$8. Click the OK button. Our resulting chart now has the proper number of classes (7).

In addition to changing these features, we will probably also want to enlarge the chart for greater clarity and display meaningful axis labels. To enlarge the chart:

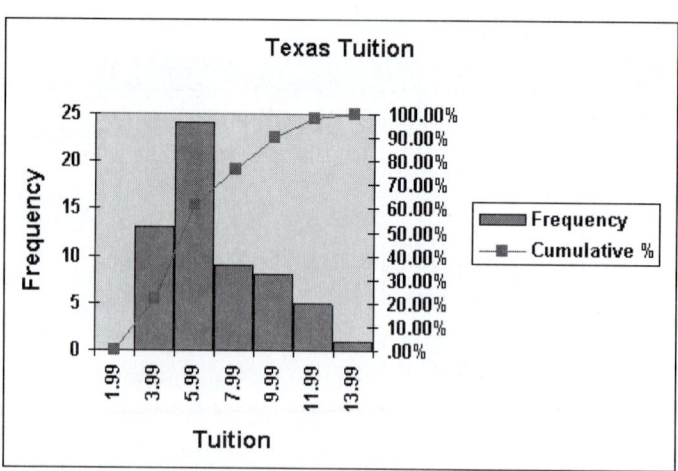

FIGURE 2.7.EXCEL Revised histogram obtained from Excel for the Texas out-of-state tuition data.

❶ First select a cell outside the chart and then single-click in the white area inside the chart box. A set of eight square handles appears on the border of the chart box.

❷ Move the mouse pointer directly over the lower left hand corner of the chart box. The mouse pointer changes to a small double-sided arrow. Drag the mouse pointer (recall that dragging requires holding down the mouse button while moving the mouse) toward cell D15. The mouse pointer changes to a simple plus sign, and as you drag it, the border of the chart box expands.

❸ While the mouse pointer is over cell D15, release the mouse button. The chart is now enlarged. Note that additional tick marks are now displayed on the *Y*-axis for each chart.

To change the label on the *X*-axis, first select a cell outside the chart box and, with the mouse pointer directly over the current *X*-axis label (Bin), double-click the mouse. A haloed border with handles appears around the word Bin. Click inside the haloed border and replace the label bin with the label Tuition.

To change the title, first select a cell outside the chart box and with the mouse pointer directly over the current title Histogram, double-click the mouse. A haloed border with handles appears around the word Histogram. (If a haloed border appears around the *X*-axis label, click one more time on the current title.) Click inside the haloed border and replace the label Histogram with the label Texas Tuition. Click outside the chart box to complete the editing of these labels. Other parts of the chart could be reformatted by double-clicking them and then selecting the appropriate Format command.

The histogram that results from these changes will be similar to the one presented in Figure 2.7.Excel.

2.7.4 Using the Chart Wizard to Obtain Frequency Polygons and Histograms

Thus far, we have used the FREQUENCY function to obtain various tables and the Data Analysis tool to obtain a variety of tables and charts. We will now discuss how the Microsoft Excel Chart Wizard can be used to develop a frequency polygon and a histogram.

The Microsoft Excel Chart Wizard contains many different types of charts that can be developed for both numerical and categorical variables. The Chart Wizard can be used by clicking on the Chart Wizard icon on the top (Standard) toolbar (it is the third icon to the left of the zoom control on the right-hand side of the toolbar) or by issuing the command Insert | Chart.

The Chart Wizard presents five linked dialog boxes, one at a time, that allow you to create a variety of charts for both numerical and categorical variables. The first dialog box asks you to provide the range for your data. The second dialog box asks you to select the type of chart desired. The third dialog box gives you the choice of many options that relate to the chart that has been chosen in the second dialog box. The fourth dialog box allows you to specify the orientation of your data and identify names for categories and data sets. The fifth dialog box allows you to format the chart by entering such things as titles and a legend. Charts created by the Chart Wizard can be further edited and formatted as will be described later in this section.

To obtain a frequency polygon from the Chart Wizard, open the workbook you developed in Section 2.7.2 or open the workbook file TEXAS-3.XLS. Note that in the design of the Calculations sheet, the upper class boundaries appear to the left of the cells containing the frequencies. We must arrange these boundaries and frequencies in this manner in order to have the Chart Wizard prepare a frequency polygon or histogram with a properly labeled X-axis. To obtain the frequency polygon from the Chart Wizard, do the following:

TEXAS-3.XLS

❶ Select Insert | Chart | As New Sheet (since we will want to place the frequency polygon on a separate sheet). The first dialog box of the Chart Wizard, asking for the range of the data to be plotted, appears.

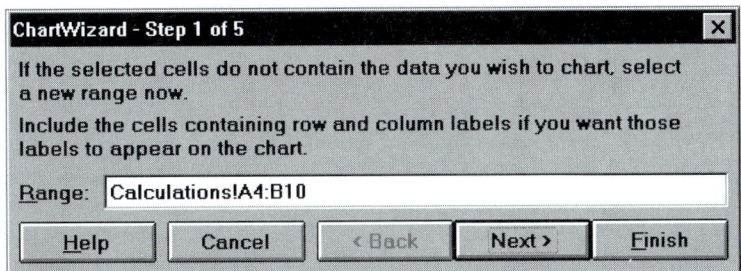

FIGURE 2.8.EXCEL   Chart Wizard steps to obtain a line chart for the Texas out-of-state tuition data. Panel A.

❷ In the Range edit box, enter Calculations!A4:B10 (Figure 2.8.Excel, Panel A). If the upper class boundaries and the frequencies do not appear in adjacent columns, enter the range of the upper class boundaries followed by a comma followed by the range of the frequencies. (For example, if the frequencies appeared in column M, rows 4–10, we would write Calculations!A4:A10,Calculations!M4:M10. Click the Next button to continue to the second dialog box.

❸ In the second dialog box, select the Line Chart choice and click the Next button. (See Panel B.)

❹ In the third dialog box, select the first choice, a line chart that contains plotted points. (See Panel C.) Click the Next button.

❺ In the fourth dialog box, select the Columns option button under the Data Series in heading. Then enter 1 in the First Columns edit box and enter 0 in the First Rows edit box. (See Panel D.) Click the Next button to move to the next dialog box.

❻ In the fifth dialog box (Panel E), you can enter a chart legend for the chart as well as titles for the axes. To obtain the frequency polygon in Figure 2.9.Excel, (a) select the No option button for Add a legend, and (b) enter Texas Tuition as the title, enter

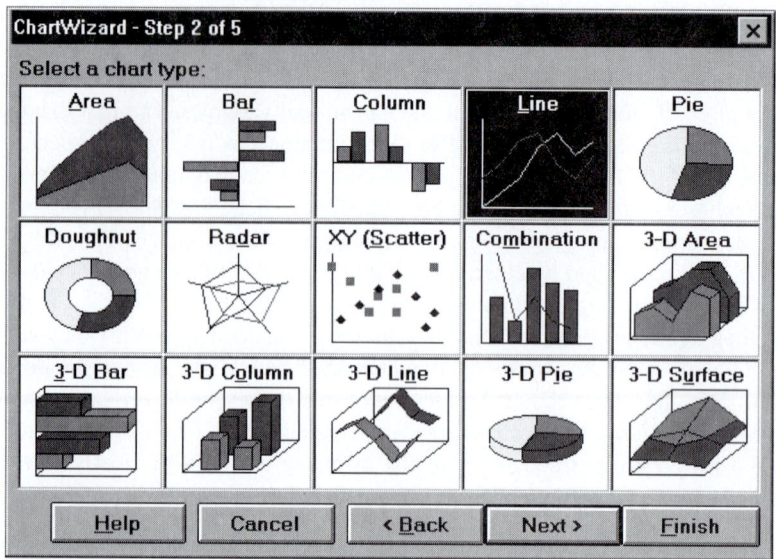

FIGURE 2.8.EXCEL   Excel Chart Wizard Step 2. Panel B.

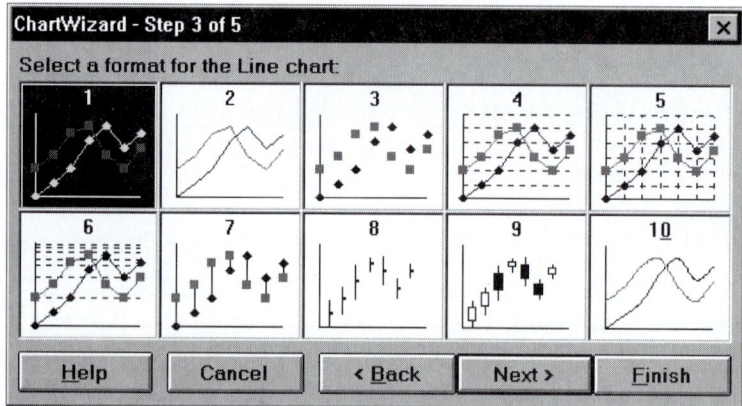

FIGURE 2.8.EXCEL   Excel Chart Wizard Step 3. Panel C.

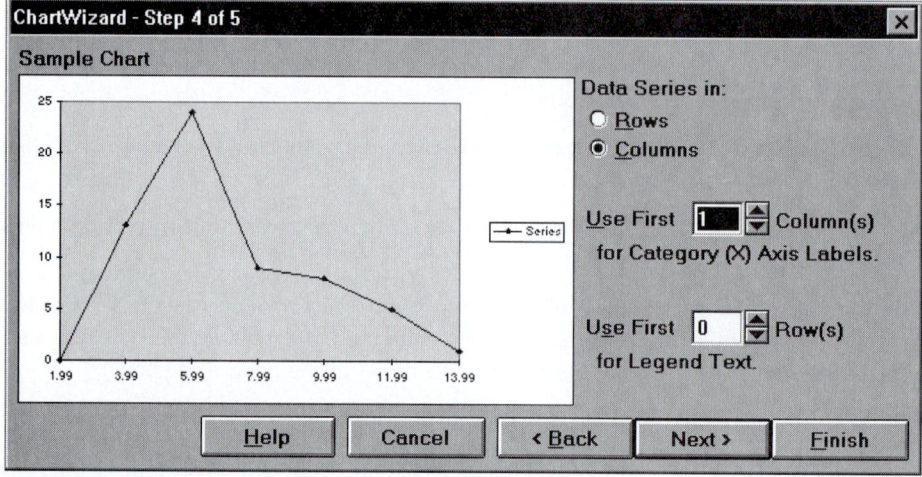

FIGURE 2.8.EXCEL   Excel Chart Wizard Step 4. Panel D.

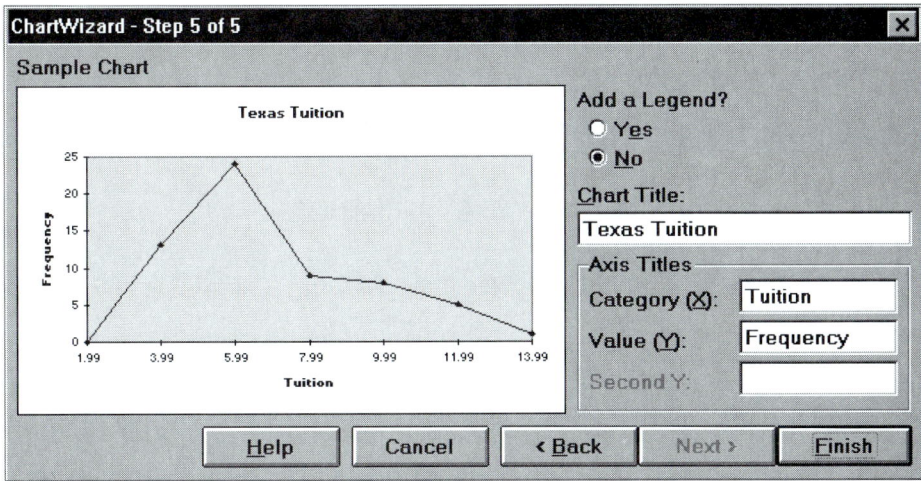

FIGURE 2.8.EXCEL   Excel Chart Wizard Step 5. Panel E.

Tuition for the category (X) edit box, and enter Frequency for the Value (Y) edit box. (See Panel E.) Click the Finish button.

Figure 2.9.Excel, Panel A, displays the frequency polygon we have obtained. Rename this sheet as Frequency Polygon.

If we examine this frequency polygon, we see that, as was the case in Figure 2.6.Excel on page 79, the category markings on the *X*-axis refer to the upper limits of the classes, not the class midpoints. To change these markings, double-click on the *X*-axis. The Format Axis dialog box appears. Select the Scale Tab and select the Value (Y) Axis crosses between categories check box. Click the OK button. Then change the class labels found in the range Calculations!A4:A10 to 1, 3, 5, 7, 9, 11, and 13. Figure 2.9.Excel, Panel B, displays the revised frequency polygon.

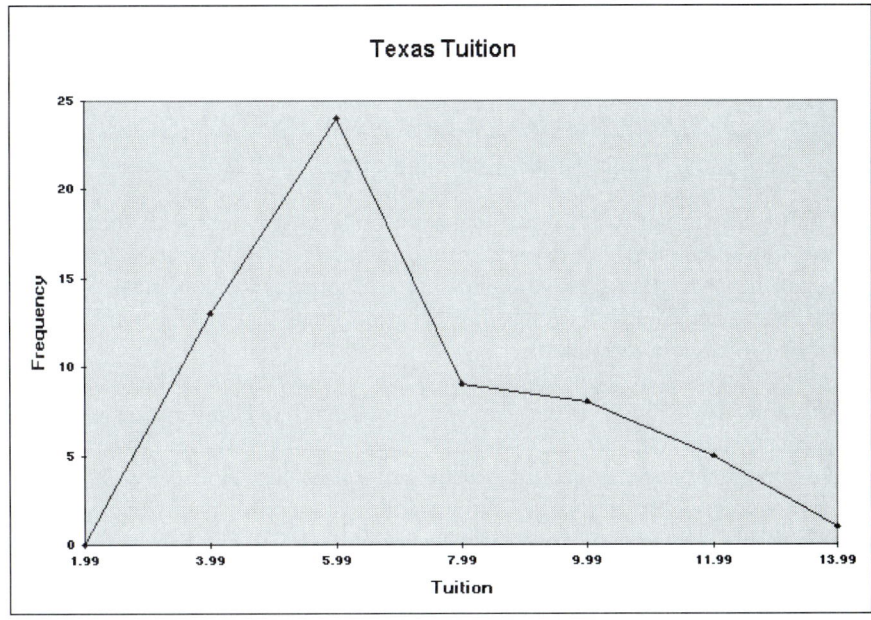

FIGURE 2.9.EXCEL   Frequency polygon for the Texas out-of-state tuition data.
Panel A.

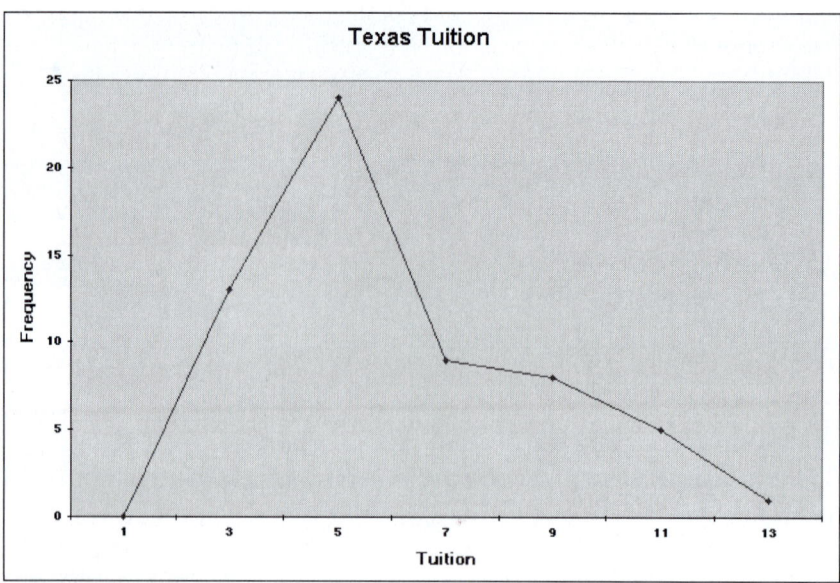

FIGURE 2.9.EXCEL   Revised frequency polygon obtained from Excel for the Texas out-of-state tuition data. Panel B.

In addition, if we wanted to obtain a percentage polygon or a cumulative percentage polygon, we would use the column of percentages or cumulative percentages instead of the column of frequencies when we define the range for the data.

To obtain a histogram from the Chart Wizard, we would follow steps similar to those described for the polygon. Steps 1 and 2 would be the same, as would steps 5 and 6. In step 3, we would select the Column Chart, while in step 4, we would select choice 8, a series of bars attached to each other.

As was the case with the polygon, if we wanted a percentage histogram, we would use the column of percentages instead of the column of frequencies.

2.8   ORGANIZING AND TABULATING CATEGORICAL DATA: THE SUMMARY TABLE

Thus far in this chapter we have learned that when collecting a large set of numerical data, the best way to examine it is first to organize and present it in appropriate tabular and chart format. Often, however, the data we collect are categorical, not numerical. Thus, in the remainder of this chapter, we will demonstrate how categorical data can be organized and presented in the form of tables and charts.

In order to do this, let us suppose once again that our analyst at the college advisory service wants to evaluate various features pertaining to colleges and universities in the state of North Carolina. Special Data Set 1 in Appendix D on page D3 displays information on the out-of-state tuition rate, type of institution, setting of school, academic calendar, and institutional focus for each of the 45 North Carolina colleges and universities. We note that the tuition rate variable is *numerical* while the other variables are all *categorical*. Earlier in this chapter we were concerned only with the former; a detailed study of the responses to the categorical variables will be undertaken here.

When dealing with categorical phenomena, the observations may be tallied into *summary tables* and then graphically displayed as either *bar charts, pie charts,* or *Pareto diagrams.*

Table 2.11 Frequency and percentage summary table pertaining to institutional focus for 45 colleges and universities in North Carolina.

Institutional Focus	Number of Schools	Percentage of Schools
National Liberal Arts Schools (NLA)	2	4.4
National Universities (NU)	4	8.9
Regional Liberal Arts Schools (RLA)	16	35.6
Regional Universities (RU)	22	48.9
Specialty Schools (SS)	1	2.2
Totals	45	100.0

Source: Data are taken from Special Data Set 1, Appendix D, page D3.

To illustrate the development of a **summary table,** let us consider the data obtained by our analyst on institutional focus. From Special Data Set 1 in Appendix D, we see that, of the 45 colleges and universities in North Carolina, 2 are classified by the College Counsel as national liberal arts schools (NLA), 4 are classified as national universities (NU), 16 are regional liberal arts schools (RLA), 22 are regional universities (RU), and 1 is a specialty school (SS). This information is presented in Table 2.11.

2.9 GRAPHING CATEGORICAL DATA: BAR AND PIE CHARTS

To express the information provided in Table 2.11 graphically, the percentage bar chart (Figure 2.10) or percentage pie chart (Figure 2.11) can be displayed.

Figure 2.10 depicts a percentage bar chart for the North Carolina institutional focus data presented in Table 2.11. In **bar charts,** each category is depicted by a bar, the length of which represents the frequency or percentage of observations falling into a category.

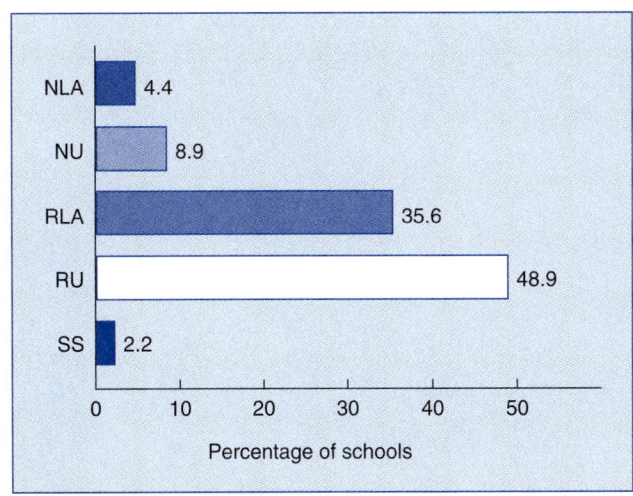

FIGURE 2.10 Percentage bar chart depicting the institutional focus of the North Carolina schools.
Source: Data are taken from Table 2.11.

To construct a bar chart, the following suggestions are made:

1. The bars should be constructed horizontally (as in Figure 2.10) when the categorized observations are the outcomes of a categorical variable. The bars should be constructed vertically when the categorized observations are the outcomes of a numerical variable.

2. All bars should have the same width (as in Figure 2.10) so as not to mislead the reader. Only the lengths should differ.

3. Spaces between bars should range from one-half the width of a bar to the width of a bar.

4. Scales and guidelines are useful aids in reading a chart and should be included. The zero point or origin should be indicated.

5. The axes of the chart should be clearly labeled.

6. Any "keys" to interpreting the chart may be included within the body of the chart or below the body of the chart.

7. Footnotes or source notes, when appropriate, are presented after the title of the chart or at the bottom edge of the chart's frame.

Figure 2.11 depicts a percentage pie chart for the North Carolina institutional focus data presented in Table 2.11.

If we had to construct a **pie chart** (when software such as Microsoft Excel was not available), we would use both the compass and the protractor—the former to draw the circle, the latter to measure off the appropriate pie sectors. Since the circle has 360°, the protractor would be used to divide up the pie based on the percentage "slices" desired. As an example, in Table 2.11, 8.9% of the North Carolina schools are classified by the College Counsel as national universities. Thus, we would multiply 360 by .089, mark off the resulting 32° with the protractor, and then connect the appropriate points to the center of the pie, forming a slice comprising 8.9% of the area of the pie.

Using this procedure with all the categories in Table 2.11 would enable us to construct the entire pie chart displayed in Figure 2.11. However, we certainly recommend that, when available, software be used for developing a pie chart.

The purpose of graphical presentation is to display data accurately and clearly. Figures 2.10 and 2.11 attempt to convey the same information with respect to institutional focus. Whether these charts succeed, however, has been a matter of much concern (see References 2–4, 11, 12). In particular, research in the human perception of graphs (Reference 4) concludes that the pie chart presents the weakest display. The bar chart is preferred to the pie chart, because it has been observed that the human eye can more accurately judge length comparisons against a fixed scale

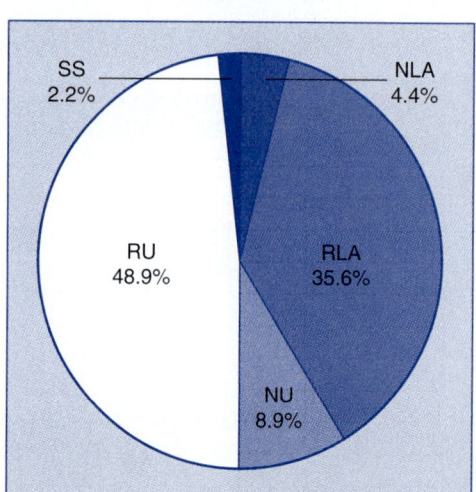

FIGURE 2.11 **Percentage pie chart depicting the institutional focus of the North Carolina schools.** *Source:* Data are taken from Table 2.11.

(as in a bar chart) than angular measures (as in a pie chart). Nevertheless, the pie chart has two distinct advantages: (1) it is aesthetically pleasing, and (2) it clearly shows that the total for all categories or slices of the pie adds to 100%. Thus, the selection of a particular chart is still highly subjective and often dependent on the aesthetic preferences of the user.

Problems for Section 2.9

Note: *To use Microsoft Excel to solve these problems, refer to Section 2.12.*

2.31 The board of directors of a large housing cooperative wishes to investigate the possibility of hiring a supervisor for an outdoor playground. All 616 households in the cooperative were polled, with each household having one vote, regardless of its size. The following data were collected:

Should the co-op hire a supervisor?	
Yes	146
No	91
Not sure	58
No response	321
Total	616

(a) Convert the data to percentages and construct
  (1) A bar chart
  (2) A pie chart
(b) Which of these charts do you prefer to use here? Why?
(c) Eliminating the "no response" group, convert the 295 responses to percentages and construct
  (1) A bar chart
  (2) A pie chart
(d) **ACTION** Based on your findings in (a) and (c), what would you recommend that the board of directors do? Write a letter to the president of the board.

2.32 The following data represent the market shares (in percent) held by manufacturers of portable, transportable, and mobile cellular phones sold during 1992:

Manufacturer	Market Share (in %)
Motorola	22
Nokia	14
Mitsubishi	10
NovAtel	9
Toshiba	8
All others	37
Total	110

Source: The New York Times, October 31, 1993, Sect. 3, p. 1.

(a) Construct a bar chart.
(b) Construct a pie chart.
(c) Which of these charts do you prefer to use here? Why?

2.33 The data at the top of page 88 represent the market share (in percent) held by manufacturers of Windows business applications software during 1992:

Manufacturer	Market Share (in %)
Aldus	4.0
Lotus	14.6
Microsoft	60.0
Software Publishing	2.9
Wordperfect	9.6
Others	8.8
Totals	99.9*

* Due to rounding.

Source: The New York Times, October 13, 1993, p. D1.

(a) Construct a bar chart.
(b) Construct a pie chart.
(c) Which of these charts do you prefer to use here? Why?

2.34 Imports to the United States from developing countries accounted for 41.4% of an estimated total of 575.9 billion dollars in the year 1993. On the other hand, exports from the United States to developing countries accounted for 40.7% of an estimated total of 459.6 billion dollars in that year. The following table presents a breakdown by country or region (in percent share) of U.S. imports and exports for the year 1993:

Country or Region	Imports into U.S. Percent Share	Exports from U.S. Percent Share
Africa	2.3	1.6
Asia (excluding Japan)	23.5	17.2
Canada	19.2	21.7
European Community	16.6	20.8
Japan	18.4	10.4
Latin America	12.9	16.8
Mideast	2.7	4.7
Other	4.4	6.8
Total	100.0	100.0

Source: The New York Times, December 19, 1993, p. F7.

(a) Construct separate bar charts for imports and exports.
(b) Construct separate pie charts for imports and exports.
(c) Which of these charts do you prefer to use here? Why?
(d) **ACTION** Analyze the data and write a memo to your economics professor based on your findings.

2.10 GRAPHING CATEGORICAL DATA: THE PARETO DIAGRAM

The **Pareto diagram** is a special type of vertical bar chart in which the categorized responses are plotted in the descending rank order of their frequencies and combined with a cumulative polygon on the same scale. The main principle behind this graphical device is its ability to sep-

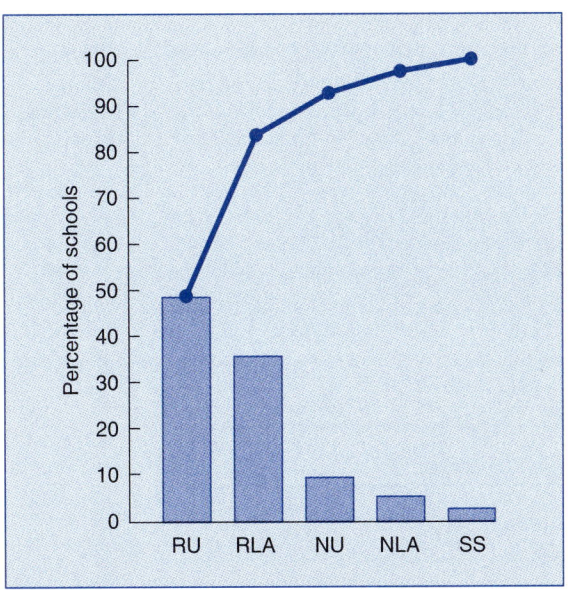

FIGURE 2.12 Pareto diagram depicting the institutional focus of the North Carolina schools.
Source: Data are taken from Table 2.11 on page 85.

arate the "vital few" from the "trivial many," enabling us to focus on the important responses. Hence the chart achieves its greatest utility when the categorical variable of interest contains many categories. The Pareto diagram is widely used in the statistical control of process and product quality (see Chapter 10).

To illustrate the Pareto diagram, we observe that in Figure 2.10, the bar chart pertaining to institutional focus presents the categories as national liberal arts schools, national universities, regional liberal arts schools, regional universities, and specialty schools. Since regional universities and regional liberal arts schools dominate the institutional focus in the state of North Carolina, a Pareto diagram may be formed by changing the ordering. Such a plot is depicted in Figure 2.12. From the lengths of the vertical bars, we observe that almost one out of every two of these schools is classified as a regional university. From the cumulative polygon, we note that 84.5% of these schools are classified as either a regional university or a regional liberal arts school.

In constructing the Pareto diagram, the vertical axis contains the percentages (from 100 on top to 0 on bottom) and the horizontal axis contains the categories of interest. The equally spaced bars must also be of equal width and, for visual impact (Reference 10), we suggest that the bars be of the same tint. The point on the cumulative percentage polygon for each category is centered at the midpoint of each respective bar. Hence, when studying a Pareto diagram, we should be focusing on two things—the magnitudes of the differences in bar lengths corresponding to adjacent descending categories and the cumulative percentages of these adjacent categories.

Problems for Section 2.10

Note: To use Microsoft Excel to solve these problems, refer to Section 2.12.

2.35 Refer to the data in Problem 2.33 on pages 87–88 regarding percent market share attained by manufacturers of Windows business applications software.
(a) Form a Pareto diagram.

(b) Which of the graphs seems to give the most visual impact—the Pareto diagram here or one of the charts drawn in (a)–(b) of Problem 2.33? Discuss.

2.36 Refer to Problem 2.34 on page 88 regarding imports and exports.
(a) Set up a table based on estimated balance of trade. That is, for each country or region, calculate the estimated value of import dollars minus export dollars—yielding either a trade deficit or a trade surplus.
(b) For the countries or regions with which the United States has a trade deficit (i.e., import dollars are more than export dollars), form a Pareto diagram.
(c) Summarize your findings.
(d) **ACTION** Write a report for your economics professor based on your findings in (c). List potential social, political, cultural, and/or economic reasons that may have led to this trade deficit.

2.37 The following data represent daily water consumption per household in a suburban water district during a recent summer:

Reason for Water Usage	Gallons per Day
Bathing and showering	99
Dish washing	13
Drinking and cooking	11
Laundering	33
Lawn watering	150
Toilet	88
Miscellaneous	20
Total	414

(a) Form a Pareto diagram.
(b) Summarize your findings.
(c) **ACTION** Since the town council is concerned about future water shortages, write a letter based on your findings in (b) pinpointing problem areas and proposing legislation that might conserve water through changes in personal habits.

2.38 The following data represent the oil production of OPEC members in December 1992, in millions of barrels per day:

Country	Daily Oil Production (in millions of barrels)
Algeria	0.77
Gabon	0.30
Indonesia	1.35
Iran	3.50
Iraq	0.55
Kuwait	1.30
Libya	1.45
Nigeria	1.90
Qatar	0.42
Saudi Arabia	8.20
United Arab Emirates	2.25
Venezuela	3.50
Total	25.49

Source: The New York Times, January 25, 1993, p. D2.

(a) Form a Pareto diagram.

(b) Summarize your findings.

(c) **ACTION▶** Write a letter to the business editor of your local newspaper on this matter.

2.39 A patient-satisfaction survey conducted for a sample of 210 individuals discharged from a large urban hospital during the month of June led to the following list of 384 complaints:

Reason for Complaint	Number
Anger with other patients/visitors	13
Failure to respond to buzzer	71
Inadequate answers to questions	38
Lateness for tests	34
Noise	28
Poor food service	117
Rudeness of staff	62
All others	21
Total	384

(a) Form a Pareto diagram.

(b) Summarize your findings.

(c) **ACTION▶** Write a memo to the chief executive officer of the hospital regarding your findings and offer suggestions for improvement.

2.40 The following table presents the number of stockholder meetings held outside the United States at which U.S. pension-fund clients of *Global Proxy Services Corporation* voted in the 1992–93 proxy season:

Country	Number of Meetings Held
Australia	49
Belgium	50
Canada	87
England	374
France	72
Germany	99
Holland	83
Hong Kong	116
Italy	115
Japan	1,249
Switzerland	61
Other	396
Total	2,751

Source: The New York Times, July 16, 1993, p. D1.

(a) Form a Pareto diagram.

(b) Summarize your findings.

(c) **ACTION▶** Write a letter to the business editor of your local newspaper on this matter.

2.11 TABULARIZING CATEGORICAL DATA USING CONTINGENCY TABLES

It is often desirable to examine the responses to two categorical variables simultaneously. For example, our analyst at the college advisory service might be interested in examining whether or not there is any pattern or relationship between type of institution (i.e., private or public) and the institutional focus. Using Special Data Set 1 in Appendix D on page D3, Table 2.12 depicts this information for all 45 colleges and universities in the state of North Carolina. Such two-way tables of cross-classification are known as **contingency tables.**

NCC&U.TXT

Table 2.12 Contingency table displaying type of institution and institutional focus of the North Carolina schools.

Type of Institution	Institutional Focus					
	NLA	NU	RLA	RU	SS	Totals
Private	2	1	16	11	0	30
Public	0	3	0	11	1	15
Totals	2	4	16	22	1	45

Source: Data are taken from Special Data Set 1, Appendix D, page D3.

To construct Table 2.12, for example, the joint responses for each of the 45 schools with respect to type of institution and institutional focus are tallied into one of the 10 possible "cells" of the table. Thus from Special Data Set 1 in Appendix D on page D3, the first school listed (Appalachian State University) is a public regional university. These joint responses were tallied into the cell composed of the second row and fourth column. The second institution (Barber Scotia College) is a private regional liberal arts school. These joint responses were tallied into the cell composed of the first row and third column. The remaining 43 joint responses were recorded in a similar manner.

In order to explore any possible pattern or relationship between type of institution and the institutional focus, it is useful to first convert these results into percentages based on

1. The overall total (i.e., the 45 colleges and universities in North Carolina)
2. The row totals (private or public)
3. The column totals [national liberal arts school (NLA), national university (NU), regional liberal arts school (RLA), regional university (RU), or specialty school (SS)]

This is accomplished in Tables 2.13, 2.14, and 2.15, respectively. Let us examine some of the many findings present in these tables. From Table 2.13, we note that

1. 66.7% of the institutions in North Carolina are private.
2. 8.9% of the institutions in North Carolina are classified as national universities.
3. 2.2% of the institutions in North Carolina are private national universities.

From Table 2.14, we note that

1. 53.3% of the private institutions are classified as regional liberal arts schools.
2. 73.3% of the public institutions are classified as regional universities.

Table 2.13 Contingency table displaying type of institution and institutional focus of the North Carolina schools (percentages based on overall total).

Type of Institution	Institutional Focus					
	NLA	NU	RLA	RU	SS	Totals
Private	4.4	2.2	35.6	24.4	0.0	66.7
Public	0.0	6.7	0.0	24.4	2.2	33.3
Totals	4.4	8.9	35.6	48.9	2.2	100.0

Source: Data are taken from Table 2.12.

Table 2.14 Contingency table displaying type of institution and institutional focus of the North Carolina schools (percentages based on row totals).

Type of Institution	Institutional Focus					
	NLA	NU	RLA	RU	SS	Totals
Private	6.7	3.3	53.3	36.7	0.0	100.0
Public	0.0	20.0	0.0	73.3	6.7	100.0
Totals	4.4	8.9	35.6	48.9	2.2	100.0

Source: Data are taken from Table 2.12.

Table 2.15 Contingency table displaying type of institution and institutional focus of the North Carolina schools (percentages based on column totals).

Type of Institution	Institutional Focus					
	NLA	NU	RLA	RU	SS	Totals
Private	100.0	25.0	100.0	50.0	0.0	66.7
Public	0.0	75.0	0.0	50.0	100.0	33.3
Totals	100.0	100.0	100.0	100.0	100.0	100.0

Source: Data are taken from Table 2.12.

And, from Table 2.15, we note that

1. 25.0% of the institutions classified as national universities are private.
2. 50.0% of the institutions classified as regional universities are public.

The tables, therefore, indicate a pattern: North Carolina institutions of higher learning comprise mainly regional universities and regional liberal arts schools—the former are evenly split between public and private schools and the latter are all private schools.

Problems for Section 2.11

2.41 In a recent study, researchers were looking at the relationship between the type of college attended and the level of job that people who graduated in 1975 held at the time of the study. The researchers examined only graduates who went into industry. The cross-tabulation of the data is as follows:

| | Type of College | | |
Management Level	Ivy League	Other Private	Public
High (Sr. V-P or above)	45	62	75
Middle	231	563	962
Low	254	341	732

(a) Construct a table with either row or column percentages, depending on which you think is more informative.
(b) Interpret the results of the study.
(c) What other variable or variables might you want to know before advising someone to attend an Ivy League or other private school if he or she wants to get to the top in business?

2.42 People returning from vacations in different countries were asked how they enjoyed their vacation. Their responses were as follows:

| | Response to Country | | | |
Country	Yuck	So-So	Good	Great
England	5	32	65	45
Italy	3	12	32	43
France	8	23	28	25
Guatemala	9	12	6	2

(a) Construct a table of row percentages.
(b) What would you conclude from this study?
(c) **ACTION** Write a letter to the travel editor of your local newspaper regarding your findings.

2.43 The defeat of the incumbent, George Bush, in the 1992 presidential election was attributed to poor economic conditions and high unemployment. Suppose that a survey of 800 adults taken soon after the election resulted in the following cross-classification of financial condition with education level:

| | Education Level | | | |
Financial Conditions	H.S. Degree or Lower	Some College	College Degree or Higher	Totals
Worse off now than before	261	48	38	347
No difference	104	73	41	218
Better off now than before	65	39	131	235
Totals	430	160	210	800

(a) Construct a table of column percentages.
(b) What would you conclude from this study?
(c) **ACTION** Write a letter to your political science professor regarding your findings.

2.12 USING MICROSOFT EXCEL TO OBTAIN TABLES AND CHARTS FOR CATEGORICAL VARIABLES

2.12.1 Introduction

In Sections 2.8–2.11, we studied the data for colleges and universities in North Carolina and formed a one-way summary table for the institutional focus of schools and a two-way contingency table for the type of school and its institutional focus. Using the summary table, we developed the bar chart, the pie chart, and the Pareto diagram. In this section we will discuss how we can use Excel to obtain these tables and charts for the North Carolina data. To organize our workbook for this purpose, we need to design and add a OneWayTable sheet in a workbook that contains a Data sheet similar to the one for the Texas tuition data. Table 2.3.Excel represents the design of the Data sheet.

Note that this design reserves space for a PivotTable, which is a special type of table that Microsoft Excel creates to summarize data. If summary data is already available, category labels could be entered in column A and their corresponding frequencies in column B.

Table 2.3.Excel Design of the One-Way Tables sheet for obtaining one-way summary tables and charts.

	A	B	C	D	E	F
1			One-Way Summary Table for N.C. schools			
2						
3	Reserved for					
4			Focus	Frequency	Percentage	Cumulative Percentage
5			=$A$5	=$B$5	=D5/$B$10	=E5
6			.	.	.	F5+E6
7			.	.	.	.
8			.	.	.	.
9			=$A$9	=$B$9	=D9/$B$10	F8+E9
10	Pivot Table					

2.12.2 Using the PivotTable Wizard to Obtain One-Way Summary Tables and Contingency Tables

The PivotTable is a type of table that summarizes information about different variables (called *fields* in Microsoft Excel). The results can be provided in various forms including counts, averages, or totals.

Selecting the command Data | PivotTable begins the PivotTable Wizard. Using this Wizard, we specify the variables of interest, the organization, and the type of data to be summarized. In this section, in developing one- and two-way tables for categorical variables, we will be interested in obtaining a count of the number of observations in a group or in a cell of a two-way contingency table.

To create a summary table for the institutional focus of North Carolina schools similar to Table 2.11 on page 85, open the NC-1.XLS file. Observe that the first row in the Data sheet contains the column headings Tuition, Type, Setting, Calendar, and Focus. Providing column headings allows the PivotTable Wizard to form an on-screen button for each variable.

NC-1.XLS

❶ Since we want the output of the PivotTable to appear on a separate sheet, click on Insert | Worksheet, rename the new sheet as OneWayTable, and click on the Data Sheet tab to return to the Data sheet.

❷ Select the command Data | PivotTable. The first of four PivotTable Wizard dialog boxes appears. (See Figure 2.10.Excel, Panel A.) Click on the Microsoft Excel List or Database button and click the Next button to move to the next dialog box.

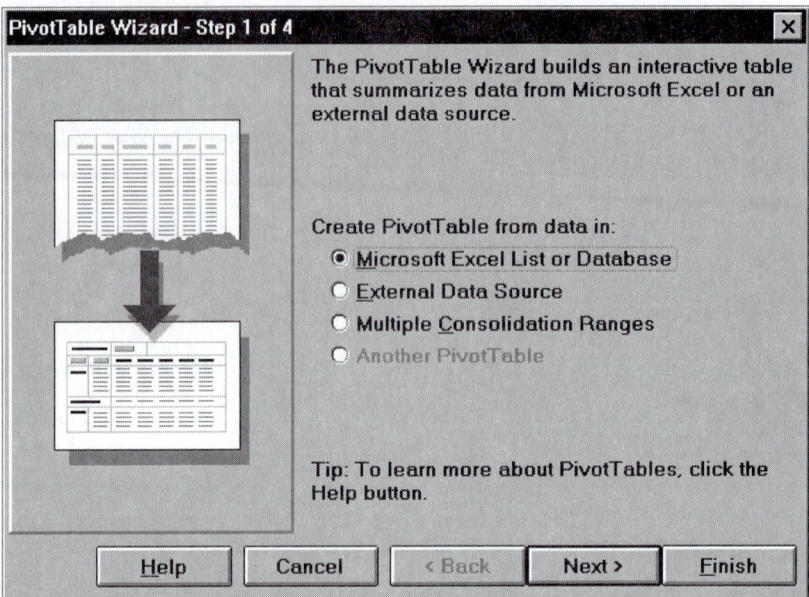

FIGURE 2.10.EXCEL PivotTable Wizard steps for the North Carolina institutional focus data. Step 1. Panel A.

❸ This second dialog box (Panel B) asks you to enter the range of your data. Enter Data!B1:F46 and click the Next button.

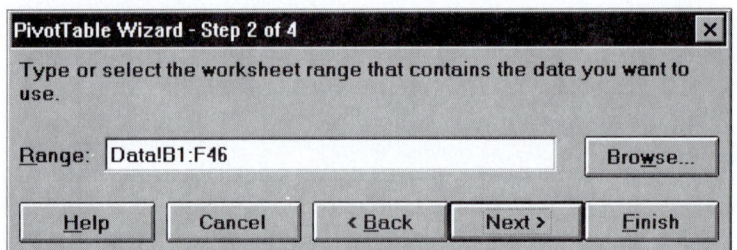

FIGURE 2.10.EXCEL PivotTable Wizard Step 2. Panel B.

❹ The third of four PivotTable Wizard dialog boxes, illustrated in Panel C, allows you to indicate the variables to be included in the PivotTable and the type of data to be analyzed. On the right side of the dialog box appear draggable labels containing the column headings for the five variables of the data sheet, tuition, type, setting, calendar, and focus. Since we wish to obtain a summary table for the variable Focus, we need to drag the Focus label and drop it (by releasing the mouse button) in the rectangular box named Row. Once you have dragged and dropped, you will observe that Focus

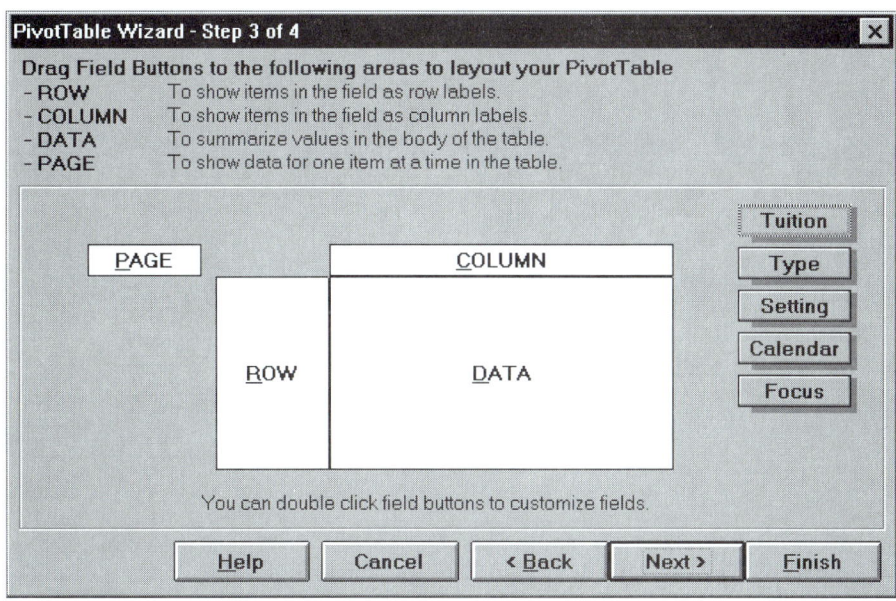

FIGURE 2.10.EXCEL   PivotTable Wizard Step 3. Panel C.

appears at the top of the Row box. Now, to obtain a count of the number of schools in each class, we select Focus again and drag and drop it to the Data area. The name in the box titled Data should change to Count of Focus. If it does not, double-click on the name indicated. This will present you with a PivotTable Field dialog box. In the Summarize by list box, select Count and then click the Options button to display the extended PivotTable Field dialog box. In the Show Data as drop down list box shown

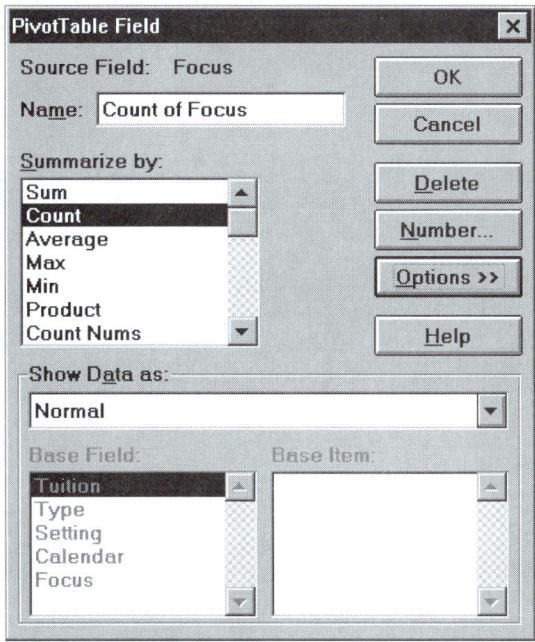

FIGURE 2.10.EXCEL   Extended PivotTable
Field dialog box. Panel D.

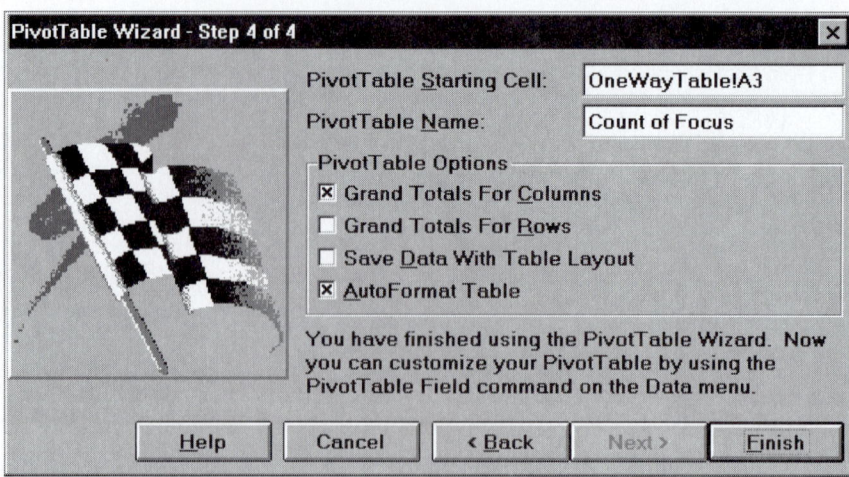

FIGURE 2.10.EXCEL PivotTable Wizard Step 4. Panel E.

in Figure 2.10.Excel, Panel D, select Normal to obtain frequencies. Click the OK button to return to dialog box 3 of the PivotTable Wizard and then click the Next button. The fourth dialog box appears.

❺ The fourth dialog box in this Wizard allows you to place and format your table. You enter OneWayTable!A3 in the PivotTable Starting Cell edit box and then enter Count of Focus in the PivotTable Name edit box. Continue by selecting the Grand Total for Columns check box to verify the count of the sample size and then the AutoFormat Table check box. (See Panel E.) Click the Finish button to have Microsoft Excel produce a PivotTable similar to the one illustrated in Figure 2.11.Excel.

	A	B
1		
2		
3	Count of Focus	
4	Focus	Total
5	NLA	2
6	NU	4
7	RLA	16
8	RU	22
9	SS	1
10	Grand Total	45

FIGURE 2.11.EXCEL Summary table of North Carolina institutional focus data.

Having used the PivotTable Wizard for the one-way summary table, we can use the Wizard a second time to obtain a two-way contingency table such as Table 2.11, on page 85, which cross-classifies the type of institution with the focus of the institution. Repeat steps 1–3 as before, this time renaming the new sheet TwoWayTable. Continue with the following steps 4 and 5:

❹ In the third PivotTable Wizard dialog box, since we wish to obtain a two-way contingency table for the variables Type (in the rows) and Focus (in the columns), we need to select and drag and drop the Type label in the row box on the left side of the dialog

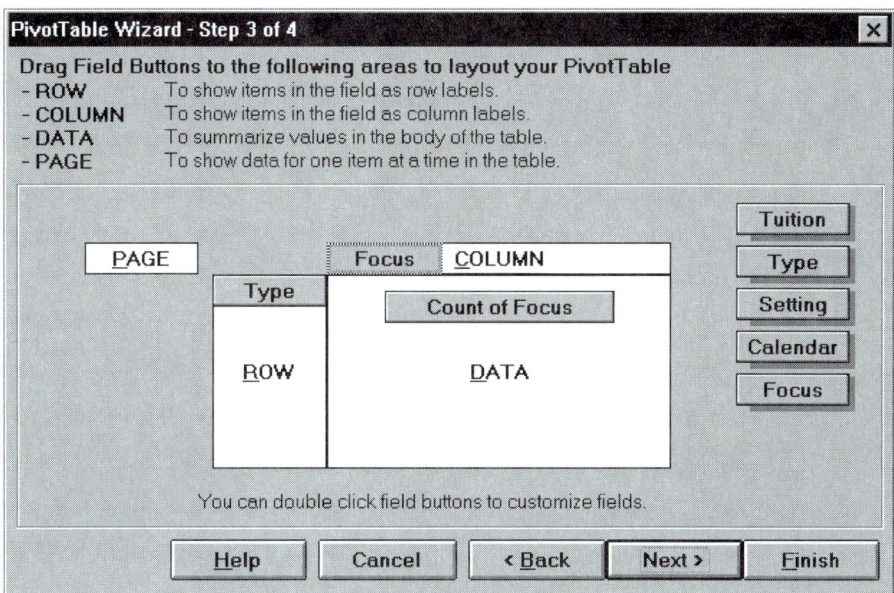

FIGURE 2.12.EXCEL   Complete PivotTable Wizard step 3 dialog box for the North Carolina institutional focus data.

box and select and drag and drop the Focus label in the Column area. Select the Focus label a second time and drag and drop it to the Data area. (See Figure 2.12.Excel.) Click the Next button. The fourth dialog box appears.

⑤ In the fourth and final dialog box (Panel E), enter TwoWayTable!A3 in the PivotTable Starting Cell edit box and then enter Type X Focus in the PivotTable Name edit box. Continue by selecting the Grand Total for Columns check box and the Grand Total for Rows check box to verify the count of the sample size and then the AutoFormat Table check box. Click the Finish button to have Microsoft Excel produce a PivotTable similar to the one illustrated in Figure 2.13.Excel.

Now suppose we wish to obtain contingency tables similar to Tables 2.13, 2.14, and 2.15 on page 93 that express the results in terms of total percentages, row percentages, and column percentages. Then in step 3 of the PivotTable Wizard from the Show Data as drop-down list box in the extended PivotTable Field dialog box, we need to select % of Total, % of Row Total, or % of Column Total, respectively.

	A3	↓		Count of Focus				
	A	**B**	**C**	**D**	**E**	**F**	**G**	
1								
2								
3	Count of Focus	Focus						
4	Type	NLA	NU	RLA	RU	SS	Grand Total	
5	Private	2	1	16	11	0	30	
6	Public	0	3	0	11	1	15	
7	Grand Total	2	4	16	22	1	45	

FIGURE 2.13.EXCEL   Contingency table obtained from Excel for the North Carolina institutional focus data.

2.12.3 Using the Chart Wizard to Obtain Bar Charts, Pie Charts, and Pareto Diagrams

In Section 2.7.4, we used the line chart option of the Chart Wizard to obtain a frequency polygon for the Texas out-of-state tuition rates. Now we will use the Chart Wizard in conjunction with the output of the one-way PivotTable to obtain a bar chart, pie chart, and Pareto diagram for the North Carolina institutional focus data summarized in Table 2.11 on page 85 and in Figure 2.11.Excel. To obtain these charts from the Chart Wizard, open the workbook file you developed in Section 2.12.2 or open the file NC-2.XLS and do the following:

NC-2.XLS ❶ Select Insert | Chart | As New Sheet (since we will want to place the bar chart on a separate sheet). The first dialog box of the Chart Wizard, asking for the range of the data to be plotted, appears.

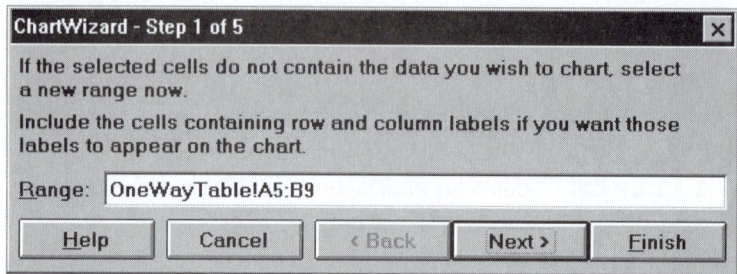

FIGURE 2.14.EXCEL   Chart Wizard steps for obtaining a bar chart for the North Carolina institutional focus data. Step 1. Panel A.

❷ In the Range edit box, enter OneWay Table!A5:B9. (See Figure 2.14.Excel, Panel A.) Note that to obtain a chart with a properly labeled X-axis, the cells containing the class categories must be located in a column to the left of the cells containing the frequencies (as is the case with this PivotTable). Click the Next button to continue to the second dialog box.

❸ In the second dialog box select the Bar Chart choice. (See Panel B.) Click the Next button.

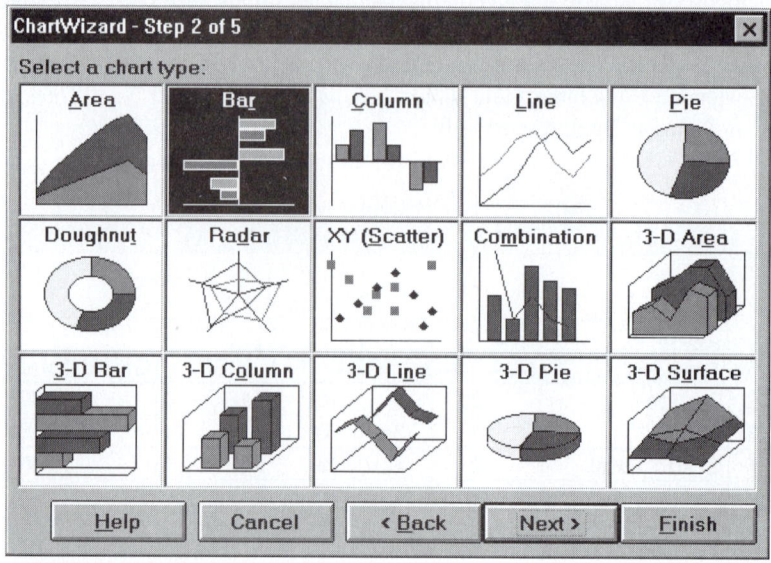

FIGURE 2.14.EXCEL   Step 2. Panel B.

④ In the third dialog box select the third choice, a bar chart that has bars going out to the right from the vertical axis. (See Panel C.) Click the Next button.

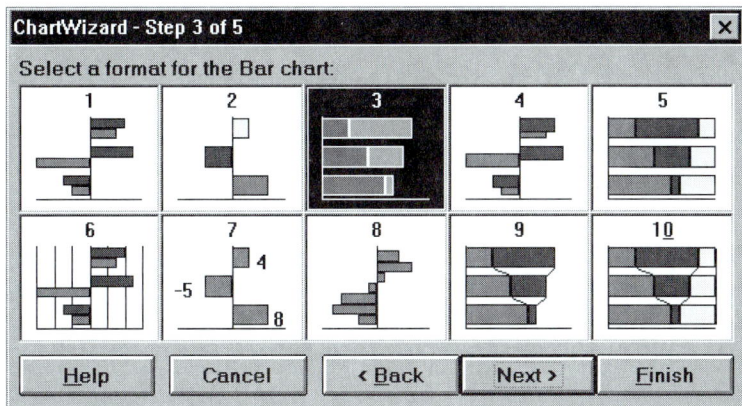

FIGURE 2.14.EXCEL   Step 3. Panel C.

⑤ In the fourth dialog box select the Columns option button under the Data Series in heading. Then enter 1 in the First Columns edit box and enter 0 in the First Rows edit box. (See Figure 2.14.Excel, Panel D.) Click the Next button to move to the next dialog box.

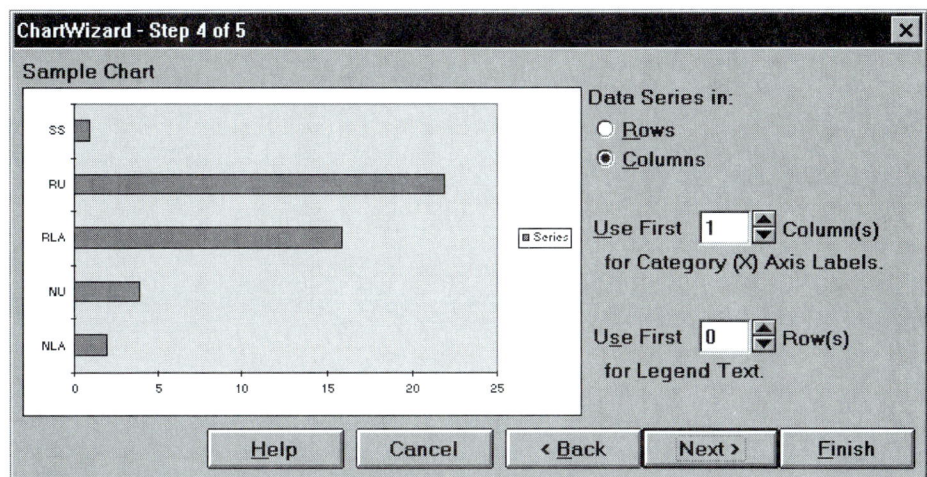

FIGURE 2.14.EXCEL   Step 4. Panel D.

⑥ In the fifth dialog box you can enter a chart legend for the chart as well as titles for the axes. To obtain the bar chart in Figure 2.15.Excel, select the No option button for Add a Legend and enter the following: N.C. Schools Focus as the title, Focus for the Category (X) edit box, and Frequency for the Value (Y) edit box. (See Panel E.) Click the Finish button.

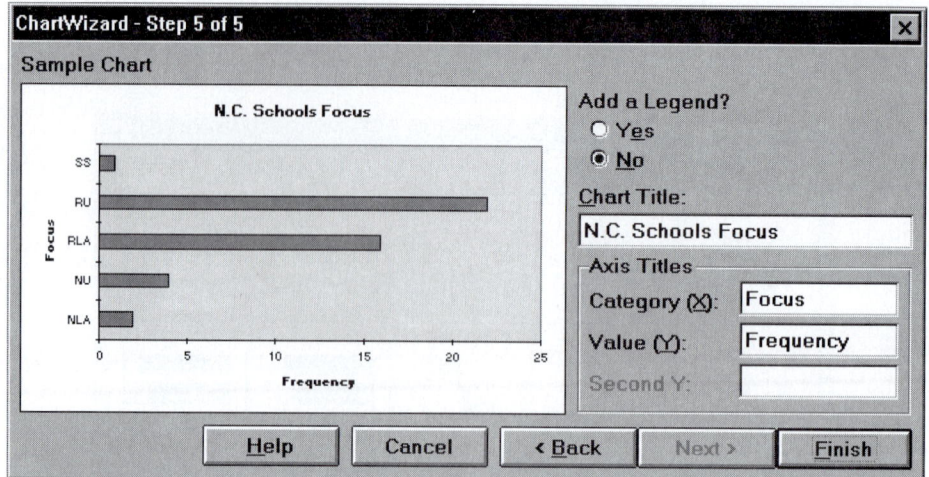

FIGURE 2.14.EXCEL Step 5. Panel E.

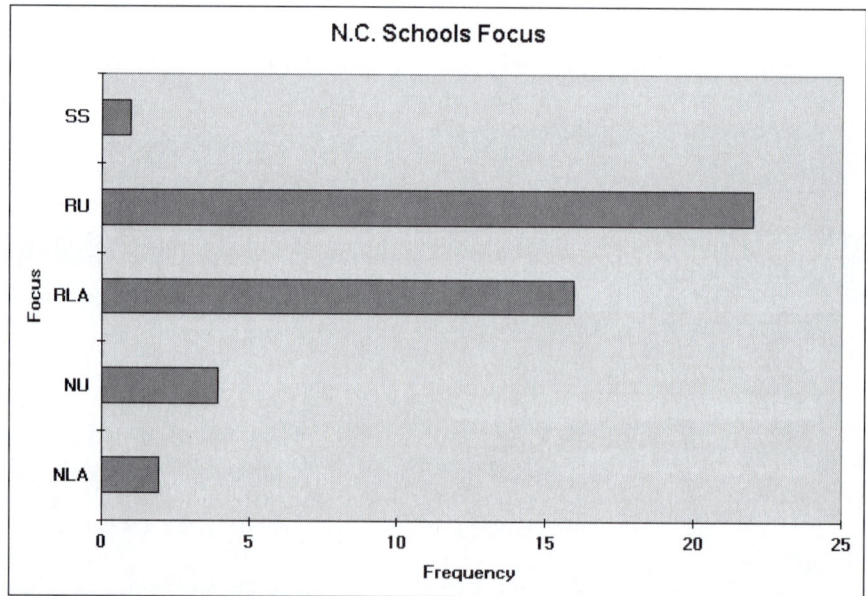

FIGURE 2.15.EXCEL Bar chart obtained from Excel for the North Carolina institutional focus data.

Figure 2.15.Excel displays the bar chart we have obtained. Rename this sheet as BarChart.

Now that we have obtained the bar chart for these data, we can return to the Chart Wizard to obtain a pie chart. Repeat steps 1 and 2 of the previous description for the bar chart and continue with steps 3–6 as follows:

❸ Click the chart titled Pie, which is the last chart on the right in the first row (Figure 2.14.Excel, Panel B). Then click the Next button.

❹ Click on Option 7, which shows the percentages corresponding to each section of the pie. Then click the Next button.

❺ Click the Columns button for Data Series. Since the categories are in the first column, enter 1 for First Columns and enter 0 for First Rows and then click the Next button.

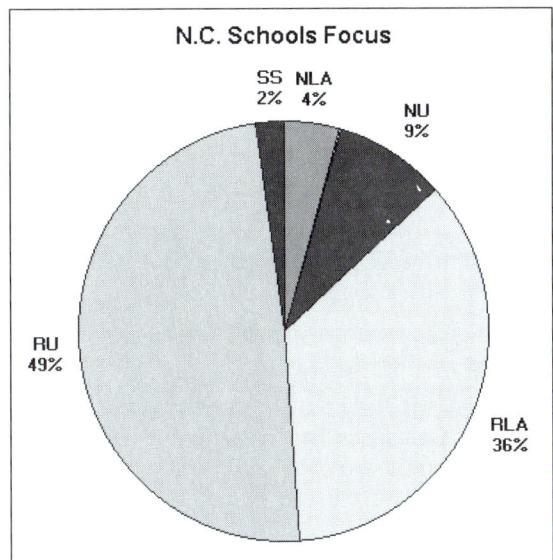

FIGURE 2.16.EXCEL Pie chart obtained from Excel for the North Carolina institutional focus data.

❻ Click the No button for Add a Legend. Select the Chart Title edit box and enter N. C. Schools Focus. Click on the Finish button.

Figure 2.16.Excel displays the pie chart we have obtained. Change the name of this sheet to Pie Chart.

Now that we have obtained the bar chart and the pie chart for these data, we would like to also develop a Pareto diagram. The Pareto diagram is an option that is not available in current versions of the Chart Wizard, so we need to use Excel formulas to develop the cumulative percentages and then use the combination chart selection on the Chart Wizard.

Assuming that we have the summary PivotTable shown in Figure 2.11.Excel, we need to copy the results of the PivotTable in columns A and B to columns C and D to be able to properly sort the categories in order using the Sort command. Then we sort the categories and frequencies in columns C and D and obtain the percentages for each category in descending order in column E and the cumulative percentages in column F. This may be accomplished by implementing the design of the OneWay Tables sheet shown in Table 2.3.Excel on page 95 and doing the following:

❶ First copy the contents of the PivotTable to columns C and D, making sure to use absolute addresses.

❷ Now we need to sort the categories in descending order according to the frequency in each category. Select the range C4:D9, and select the command Data | Sort. The Sort dialog box appears. In the Sort By drop-down list box, select Frequency and then select the Descending option button. Then click the OK button.

❸ Enter the formula =D5/$B$10 in cell E5 and copy this formula to cells E6 through E9. To change the results to percentage format, select the range E5:E9 and click the % button and then the Increase Decimal button twice (to get two decimal places).

❹ To obtain the cumulative percentage distribution, enter the formula =E5 in cell F5 and then enter =F5+E6 into cell F6. Copy the contents of cell F6 to cells F7:F9.

Now that we have obtained the percentage in each class and the cumulative percentages, we can use the Chart Wizard to obtain the Pareto diagram as we did with the bar chart and the pie chart. With the OneWayTable sheet active, select Insert | Chart | As New Sheet, and enter the following at each step of the Chart Wizard:

Wizard Step 1. Enter the range of the data as C5:C9,E5:F9. (The comma is necessary here since the data are located in two separate *nonadjacent* areas.)

Wizard Step 2. Select the Combination chart. This chart will provide a bar chart and a line chart.

Wizard Step 3. Select the second choice that will plot a separate bar and line chart with two Y-axes.

Wizard Step 4. Select the Columns option button for Data Series, enter 1 in the First Columns edit box, and enter 0 in the First Rows edit box.

Wizard Step 5. Select the No option button for Add a Legend and enter Pareto Diagram as the Chart Title and Focus in the Category X edit box.

Figure 2.17.Excel illustrates the frequency and cumulative distributions along with the Pareto diagram. Rename this sheet as Pareto.

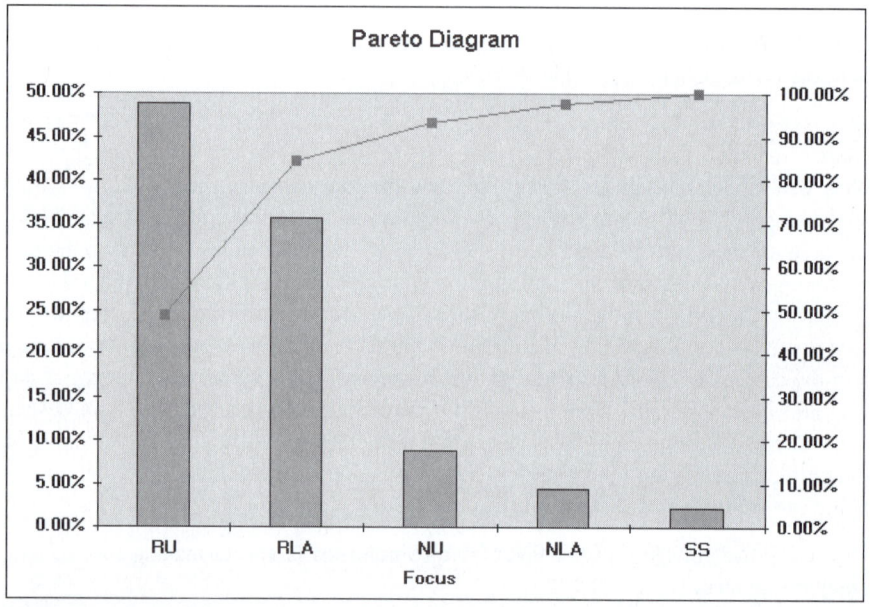

FIGURE 2.17.EXCEL Pareto diagram obtained from Excel for the North Carolina institutional focus data.

2.13 PROPER TABULAR AND CHART PRESENTATION AND ETHICAL ISSUES

To this point we have studied how a collected set of data is presented in tabular and chart form. If our analysis is to be enhanced by a visual display of data, it is essential that the tables and charts be presented clearly and carefully. Tabular frills and other **"chart junk"** must be eliminated so as not to cloud the message given by the data with unnecessary adornments (References 7, 12, 13, and 15). In addition to eliminating chart junk when we display charts, we must also avoid common errors that distort our visual impression (References 6, 9, and 10). Two such errors are

1. Failing to compare two or more sets of data on a relative basis
2. Failing to indicate the zero point at the bottom of the vertical axis

2.13.1 Failing to Compare Data Sets on a Relative Basis

In Section 2.4, we demonstrated why it is necessary to compare two or more sets of data on a relative basis, and Figures 2.7 (page 69) and 2.9 (page 73) respectively displayed the proper percentage polygons and percentage ogives comparing the out-of-state tuition rates for 60 Texas schools and 45 North Carolina schools. Using frequency counts rather than percentages or proportions would be misleading. To see this, look at Figures 2.13 and 2.14 where frequency polygons and frequency ogives improperly "compare" the Texas and North Carolina out-of-state tuition rates. In addition, to dramatize the visual distortion, we include the out-of-state tuition rates charged by the 90 Pennsylvania schools (see Special Data Set 1 of Appendix D on pages D4–D5). As can be seen from Figures 2.13 and 2.14, the frequency polygons and frequency ogives for the Texas and North Carolina schools are overwhelmed by those for the Pennsylvania schools, and no meaningful comparisons can be made from such distorted charts.

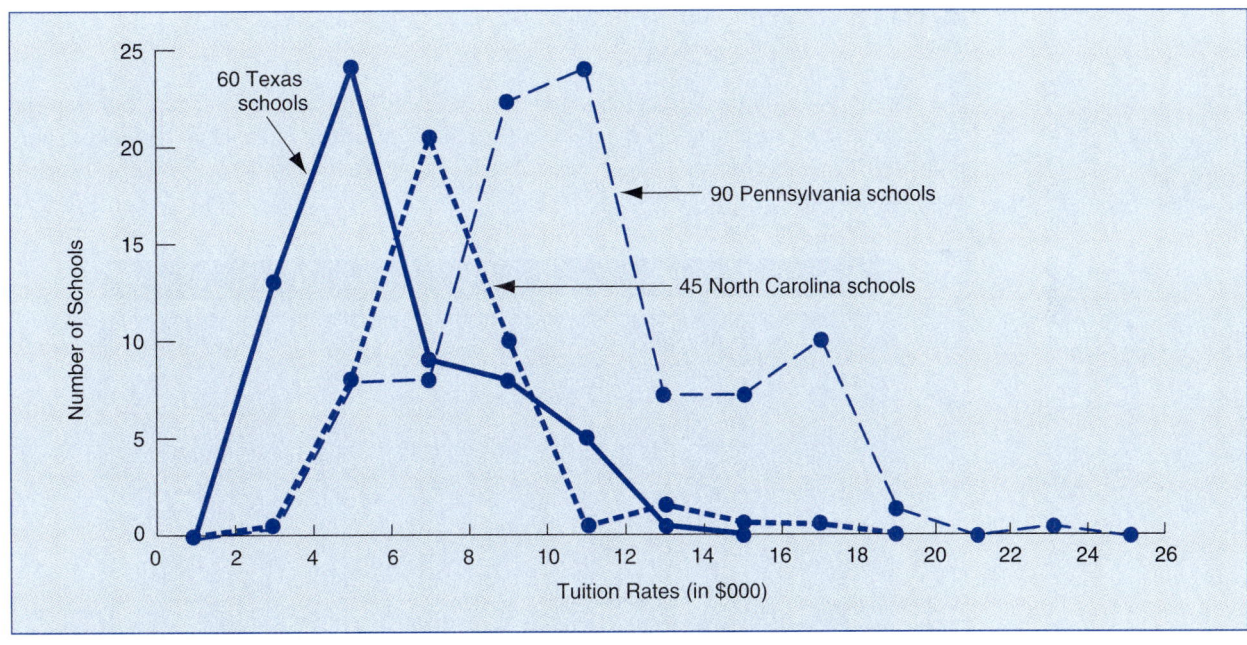

FIGURE 2.13 **"Improper" frequency polygons for the out-of-state tuition rates for schools in Texas, North Carolina, and Pennsylvania.** *Source:* Data are taken from Tables 2.3, 2.7, and "America's Best Colleges, 1994 College Guide," *U.S. News & World Report,* extracted from College Counsel 1993 of Natick, Mass. Reprinted by special permission, *U.S. News & World Report,* © 1993 by *U.S. News & World Report* and by College Counsel.

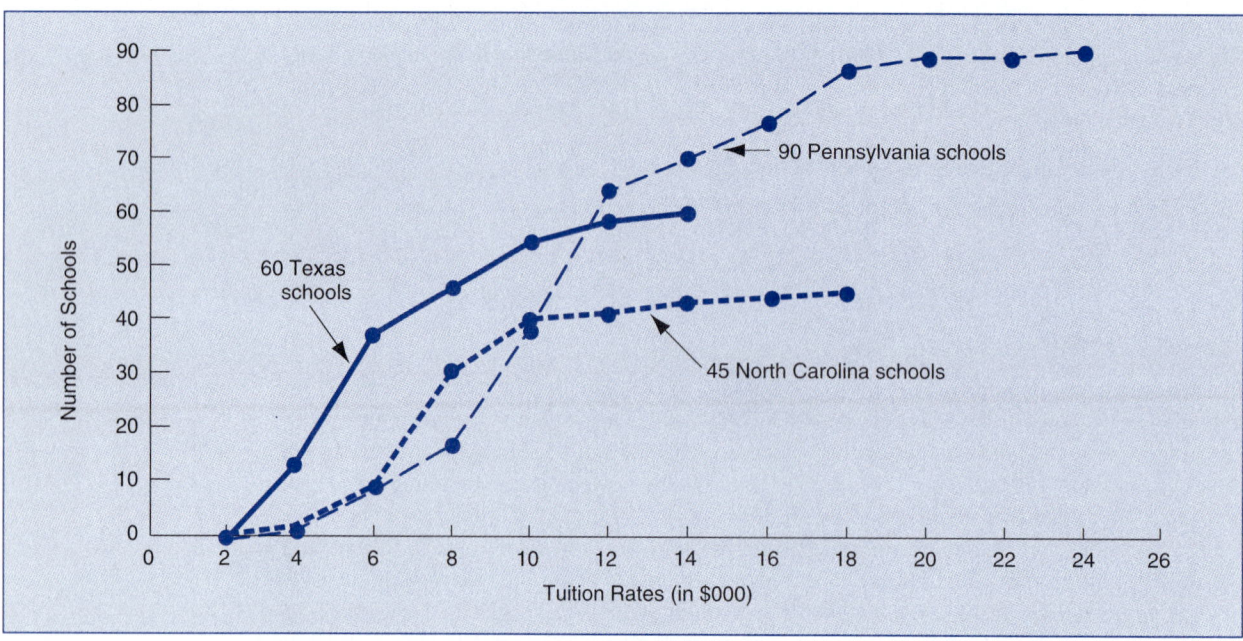

FIGURE 2.14 **"Improper" cumulative frequency polygons for the out-of-state tuition rates for schools in Texas, North Carolina, and Pennsylvania.** *Source:* Data are taken from Tables 2.3, 2.7, and "America's Best Colleges, 1994 College Guide," *U.S. News & World Report*, extracted from College Counsel 1993 of Natick, Mass. Reprinted by special permission, *U.S. News & World Report*, © 1993 by *U.S. News & World Report* and by College Counsel.

2.13.2 Failing to Indicate the Zero Point on the Vertical Axis

The starting point on the vertical axis must be indicated with a zero so as not to distort the visual impression regarding the magnitude of changes occurring in the chart. By taking only a slice of the vertical axis, such changes can be exaggerated. Figure 2.15 displays such a visual distortion.

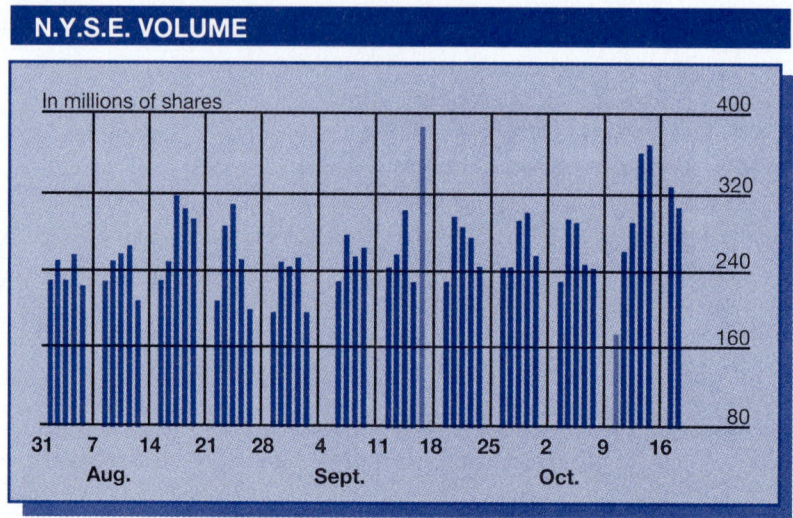

FIGURE 2.15 **"Improper" display of New York Stock Exchange sales volume (in millions of shares traded) over time.** *Source:The New York Times*, October 20, 1993, p. D7.

Type of Drink	Market Share (in %)
Beer	12
Carbonated soft drinks	25
Coffee	11
Juice	6
Milk	15
Tap water	19
Other	12
Total	100

(a) For the data on market share of all types of drinks, construct
   (1) A bar chart
   (2) A pie chart
(b) Which chart do you prefer for purposes of presentation? Why?
(c) For the data on carbonated soft drink market share, develop the appropriate graph to pinpoint the "vital few" from the "trivial many."
(d) Analyze the data and summarize your findings.
(e) **ACTION** Write a letter to the food editor of your local newspaper explaining your findings.
(f) **(Class Project)** Let each student in the class respond to the question: "Which type of carbonated soft drink do you most prefer?" so that the teacher may tally the results into a summary table on the blackboard.
   (1) Convert the data to percentages and construct a Pareto diagram.
   (2) Compare and contrast the findings from the class with those obtained nationally based on market shares. What do you conclude? Discuss.

2.53 The following data represent the global market sales of all products manufactured by Motorola, Inc., in 1992 and a stratification of its net sales by business segment:

Region	Market Sales (in %)
Asia-Pacific	15
Europe	21
Japan	7
United States	48
Other	9
Total	100

Business Segment	Net Sales (in %)
Communications	29
General systems	26
Government electronics	5
Information systems	4
Semiconductor	32
Other	4
Total	100

Source: The New York Times, October 31, 1993, Sect. 3, p. 6.

S-2-53.XLS

(a) For the data on global market sales of all products, construct
   (1) A bar chart
   (2) A pie chart
(b) Which chart do you prefer for purposes of presentation? Why?
(c) For the data on net sales by business segment, develop the appropriate graph to pinpoint the "vital few" from the "trivial many."
(d) Analyze the data and summarize your findings.
(e) **ACTION** Write a letter to your marketing professor explaining your findings.

2.54 The following table provides a percentage breakdown of where personal computers were sold in 1987 and in 1993:

Type	Percentage of Sales	
	1987	1993
Direct response	0	14
Direct sellers	17	4
Mail order	4	3
Mass merchants	3	8
Superstores	0	6
Trade dealers	60	44
Value-added resellers	11	13
Other	5	8
Totals	100	100

Source: The New York Times, May 30, 1993, p. F5.

(a) For each year construct an appropriate graph and analyze the data.
(b) **ACTION** Write a letter to your marketing professor discussing the implications of your analysis of these shifting trends.

2.55 **(Class Project)** Let each student in the class be cross-classified based on gender (male, female) and current employment status (yes, no) with the results tallied on the blackboard.
(a) Construct a table with either row or column percentages, depending on which you think is more informative.
(b) What would you conclude from this study?
(c) What other variables would you want to know regarding employment in order to enhance your findings?

2.56 Go to the World Wide Web page for this text (http://www.prenhall.com/phbusiness) for additional exercises.

Team Projects

TP2.1 The analyst at the college advisory service requires help to finish her report regarding tuition rates charged to out-of-state residents by colleges and universities in different regions of the country. In order to meet the deadline for a presentation to the board of directors, the marketing manager decides to hire your group, the _____ Corporation, to assist the analyst in her endeavors. Given Special Data Set 1 of Appendix D on pages D4–D5 regarding the out-of-state tuition rates in the Pennsylvania schools, the _____ Corporation is ready to

(a) Outline how the group members will proceed with their tasks.

(b) Form the frequency and percentage distribution in the same table.

(c) Plot the percentage polygon.

(d) Form the cumulative percentage distribution.

(e) Plot the percentage ogive.

(f) Perform a descriptive analysis comparing the tuition rates in Pennsylvania to those in Texas and North Carolina.

(g) In addition, the analyst would like to form a separate contingency table (based on row percentages) for type of institution (private or public) and setting (rural, suburban, or urban) for each of the three states.

(h) Write and submit an executive summary, attaching all tables and charts.

(i) Prepare and deliver a 10-minute oral presentation to the marketing manager.

PENNC&U.TXT

C&U.TXT

TP2.2 A popular family magazine interested in publishing an article on the dietary virtues (or lack thereof) of ready-to-eat cereals hires your group, the _____ Corporation, to study their cost and nutritional characteristics. The theme the article is intending to present is that "ready-to-eat cereals are a quick and efficient way to get the family started each weekday." Armed with Special Data Set 2 of Appendix D on pages D6–D7 displaying useful information on 84 such cereals, the _____ Corporation is ready to

CEREAL.TXT

(a) Outline how the group members will proceed with their tasks.

(b) Form the frequency and percentage distribution in the same table.

(c) Plot the percentage polygon.

(d) Form the cumulative percentage distribution.

(e) Plot the percentage ogive.

(f) Perform a descriptive analysis.

(g) Additionally, form a contingency table cross-classifying type of ready-to-eat cereal (high fiber, moderate fiber, low fiber) with level of calories per serving (below 155, at or above 155).

(h) Write and submit an executive summary, attaching all tables and charts.

(i) Prepare and deliver a 10-minute oral presentation to the food editor of the magazine.

TP2.3 The manufacturer of well-known men's and women's fragrances is planning to develop a new product line to be marketed for the upcoming holiday season. The marketing director hires your group, the _____ Corporation, to study the characteristics of currently available fragrances so the manufacturer will be in a better position to price its newly developed product line. Armed with Special Data Set 3 of Appendix D on pages D8–D9 displaying useful information on 83 such fragrances, the _____ Corporation is ready to

FRAGRANC.TXT

(a) Outline how the group members will proceed with their tasks.

(b) Form the frequency and percentage distribution in the same table.

(c) Plot the percentage polygon.

(d) Form the cumulative percentage distribution.

(e) Plot the percentage ogive.

(f) Perform a descriptive analysis.

(g) Additionally form a contingency table cross-classifying type of fragrance (perfume, cologne, or "other") with intensity of fragrance (very strong, strong, medium, or mild).

(h) Construct a table based on total percentages.

(i) Construct a table based on row percentages.

(j) Construct a table based on column percentages.

(k) Repeat (g)–(j) for women's fragrances only.

(l) Repeat (g)–(j) for men's fragrances only.

(m) Compare and contrast the results in (k) and (l).

(n) Write and submit an executive summary, attaching all tables and charts.

(o) Prepare and deliver a 10-minute oral presentation to the marketing director.

Endnote

1. The diskette that accompanies this text includes a Visual Basic for Applications module (STEMLEAF.XLS) that aids in the creation of a stem-and-leaf display.

References

1. Chambers, J. M., W. S. Cleveland, B. Kleiner, and P. A. Tukey, *Graphical Methods for Data Analysis* (Boston, MA: Duxbury Press, 1983).

2. Cleveland, W. S., "Graphs in Scientific Publications," *The American Statistician*, Vol. 38 (November 1984), pp. 261–269.

3. Cleveland, W. S., "Graphical Methods for Data Presentation: Full Scale Breaks, Dot Charts, and Multibased Logging," *The American Statistician*, Vol. 38 (November 1984), pp. 270–280.

4. Cleveland, W. S., and R. McGill, "Graphical Perception: Theory, Experimentation, and Application to the Development of Graphical Methods," *Journal of the American Statistical Association*, Vol. 79 (September 1984), pp. 531–554.

5. Cobb Group, *Running Microsoft Excel 5* (Redmond, WA: Microsoft Press, 1994).

6. Croxton, F., D. Cowden, and S. Klein, *Applied General Statistics,* 3d ed. (Englewood Cliffs, NJ: Prentice-Hall, 1967).

7. Ehrenberg, A. S. C., "Rudiments of Numeracy," *Journal of the Royal Statistical Society*, Series A, Vol. 140 (1977), pp. 277–297.

8. *Microsoft EXCEL Version 7* (Redmond, WA: Microsoft Press, 1996).

9. Huff, D., *How to Lie with Statistics* (New York: W. W. Norton, 1954).

10. Kimble, G. A., *How to Use (and Misuse) Statistics* (Englewood Cliffs, NJ: Prentice-Hall, 1978).

11. Tufte, E. R., *The Visual Display of Quantitative Information* (Cheshire, CT: Graphics Press, 1983).

12. Tufte, E. R., *Envisioning Information* (Cheshire, CT: Graphics Press, 1990).

13. Tukey, J., *Exploratory Data Analysis* (Reading, MA: Addison-Wesley, 1977).

14. Velleman, P. F., and D. C. Hoaglin, *Applications, Basics, and Computing of Exploratory Data Analysis* (Boston, MA: Duxbury Press, 1981).

15. Wainer, H., "How to Display Data Badly," *The American Statistician*, Vol. 38 (May 1984), pp. 137–147.

chapter 11

Simple Linear Regression and Correlation

CHAPTER OBJECTIVE

To develop the simple linear regression model as a means of using one variable to predict another variable and to study correlation as a measure of the strength of the association between two variables.

11.1 INTRODUCTION

In previous chapters we focused primarily on a single numerical response variable, such as the amount of out-of-state tuition. We studied various measures of statistical description (see Chapter 3) and applied different techniques of statistical inference to make estimates and draw conclusions about our numerical response variable (see Chapters 6–8). In this and the following chapter we will concern ourselves with problems involving two or more numerical variables as a means of viewing the relationships that exist between them. Two techniques will be discussed: regression and correlation.

Regression analysis is used primarily for the purpose of prediction. Our goal in regression analysis is the development of a statistical model that can be used to predict the values of a **dependent** or **response variable** based on the values of at least one **explanatory** or **independent variable**. In this chapter we shall focus on a simple regression model—one that would utilize a single numerical independent variable X to predict the numerical dependent variable Y. In Chapter 12, we shall develop a multiple regression model—one that could utilize several explanatory variables $(X_1, X_2, \ldots, X_p)$ to predict a numerical dependent variable Y.[1]

Correlation analysis, in contrast to regression, is used to measure the strength of the association between numerical variables. For example, in Section 11.8 we will determine the correlation between the price of a six-pack of soft drinks and the price of chicken in different cities. In this instance, the objective is not to use one variable to predict another, but rather to measure the strength of the association or covariation that exists between two numerical variables.

11.2 THE SCATTER DIAGRAM

Methods of regression and correlation analysis will be applied to two examples in this chapter. In the first, suppose that the management of a chain of package delivery stores would like to develop a model for predicting the weekly sales (in thousands of dollars) for individual stores. A random sample of 20 stores was selected from among all the stores in the chain. In developing such a model, many explanatory variables could be considered. However, we will begin our discussion with a simple regression model in which only one variable is used to predict the values of a dependent variable. Thus, we will develop a model to predict weekly sales (the dependent variable Y) based on the number of customers (the explanatory or independent variable X). The results for a sample of 20 stores are summarized in Table 11.1. Such data, however, can be presented in a form that is more visually interpretable.

In Chapter 2, when information concerning the out-of-state tuition rates in Texas was studied, various graphs (e.g., histograms, polygons, and ogives) were developed for data presentation. In a regression analysis involving one independent and one dependent variable, the individual values are plotted on a two-dimensional graph called a **scatter diagram**. Each value is plotted at its particular X- and Y-coordinates. The scatter diagram for the data in Table 11.1 is shown in Figure 11.1.

An examination of Figure 11.1 indicates a clearly increasing relationship between number of customers (X) and weekly sales (Y). As the number of customers increases, weekly sales increase. The exact mathematical form of the model expressing the relationship as well as methods for estimating the weekly sales for a given number of customers will be examined in subsequent sections of this chapter.

Table 11.1 Number of customers and weekly sales for a sample of 20 package delivery stores.

Stores	Customers	Sales ($000)
1	907	11.20
2	926	11.05
3	506	6.84
4	741	9.21
5	789	9.42
6	889	10.08
7	874	9.45
8	510	6.73
9	529	7.24
10	420	6.12
11	679	7.63
12	872	9.43
13	924	9.46
14	607	7.64
15	452	6.92
16	729	8.95
17	794	9.33
18	844	10.23
19	1,010	11.77
20	621	7.41

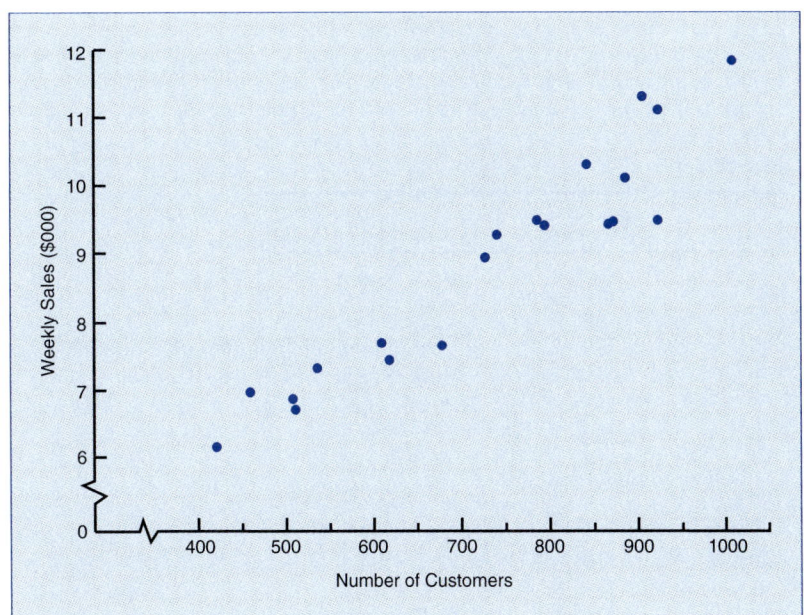

FIGURE 11.1 Scatter diagram of weekly sales and number of customers for the package delivery stores. *Source:* Data are taken from Table 11.1.

Problems for Section 11.2

Note: *The problems in this section can be solved using Microsoft Excel (see Sections 11.6.3 and 11.7).*

PETFOOD.TXT

● 11.1 The marketing manager of a large supermarket chain would like to determine the effect of shelf space on the sales of pet food. A random sample of 12 equal-sized stores is selected with the following results:

Pet food sales.

Store	Shelf Space, X (feet)	Weekly Sales, Y (hundreds of dollars)
1	5	1.6
2	5	2.2
3	5	1.4
4	10	1.9
5	10	2.4
6	10	2.6
7	15	2.3
8	15	2.7
9	15	2.8
10	20	2.6
11	20	2.9
12	20	3.1

Set up a scatter diagram.

S-SITE.XLS

11.2 Over the past 25 years, a chain of discount women's clothing stores has increased market share by increasing the number of locations in the chain. A systematic approach to site selection was never utilized. Site selection was primarily based on what was considered to be a great location or a great lease. This year, with a strategic plan for opening several new stores, the director of special projects and planning is being asked to develop an approach to forecasting annual sales for all new stores. The following represents the square footage and the annual sales (in thousands of dollars) for a sample of 14 stores in the chain:

Site selection.

Store	Square Feet	Annual Sales ($000)
1	1,726	3,681
2	1,642	3,895
3	2,816	6,653
4	5,555	9,543
5	1,292	3,418
6	2,208	5,563
7	1,313	3,660
8	1,102	2,694
9	3,151	5,468
10	1,516	2,898
11	5,161	10,674
12	4,567	7,585
13	5,841	11,760
14	3,008	4,085

Set up a scatter diagram.

11.3 A company manufacturing parts would like to develop a model to estimate the number of worker-hours required for production runs of varying lot size. A random sample of 14 production runs (2 each for lot sizes 20, 30, 40, 50, 60, 70, and 80) is selected with the following results:

WORKHRS.TXT

Production worker-hours.

Lot Size	Worker-Hours
20	50
20	55
30	73
30	67
40	87
40	95
50	108
50	112
60	128
60	135
70	148
70	160
80	170
80	162

Set up a scatter diagram.

11.4 An agronomist would like to determine the effect of a natural organic fertilizer on the yield of tomatoes. In this experiment five differing amounts of fertilizer are to be used on 10 equivalent plots of land: 0, 10, 20, 30, and 40 pounds per 100 square feet. The levels of fertilizer are randomly assigned to the plots of land with the following results:

TOMYIELD.TXT

Tomato yield.

Plot	Amount of Fertilizer, X (in pounds per 100 square feet)	Yield, Y (in pounds)
1	0	6
2	0	8
3	10	11
4	10	14
5	20	18
6	20	23
7	30	25
8	30	28
9	40	30
10	40	34

Set up a scatter diagram.

11.5 A limousine service operating from a suburban county wants to determine the length of time it would take to transport passengers from various locations to a major metropolitan airport during nonpeak times. A sample of 12 trips on a particular day during nonpeak times indicates the following as shown on page 538:

LIMO.TXT

Airport travel.

Distance (miles)	Time (minutes)
10.3	19.71
11.6	18.15
12.1	21.88
14.3	24.21
15.7	27.08
16.1	22.96
18.4	29.38
20.2	37.24
21.8	36.84
24.3	40.59
25.4	41.21
26.7	38.19

Set up a scatter diagram.

11.3 TYPES OF REGRESSION MODELS

In the scatter diagram plotted in Figure 11.1 on page 535, a rough idea of the type of relationship that exists between the variables can be observed. The nature of the relationship can take many forms, ranging from simple ones to extremely complicated mathematical functions. The simplest relationship consists of a straight-line or **linear relationship**. An example of this relationship is shown in Figure 11.2.

The straight-line (linear) model can be represented as

$$Y_i = \beta_0 + \beta_1 X_i + \epsilon_i \qquad (11.1)$$

where
$$\beta_0 = Y \text{ intercept for the population}$$
$$\beta_1 = \text{slope for the population}$$
$$\epsilon_i = \text{random error in } Y \text{ for observation } i$$

In this model, the **slope** of the line β_1 represents the expected change in Y per unit change in X. It represents the amount that Y changes (either positively or negatively) for a particular unit

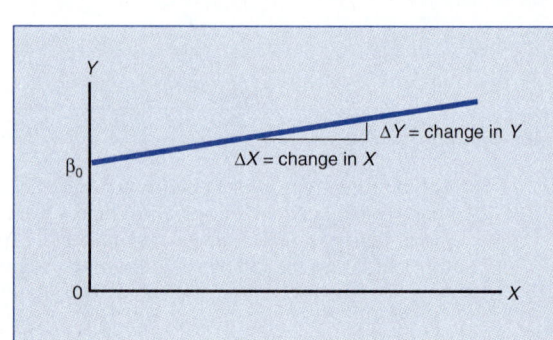

FIGURE 11.2 A positive straight-line relationship.

change in X. The **Y intercept** β_0 represents the average value of Y when X equals 0. The last component of the model, ϵ_i, represents the random error in Y for each observation i that occurs.

The selection of the proper mathematical model is influenced by the distribution of the X and Y values on the scatter diagram. This can be seen readily from an examination of Panels A–F in Figure 11.3. Thus, from Panel A, we note that the values of Y are generally increasing linearly as X increases. This panel is similar to Figure 11.1, which illustrates the relationship between number of customers and sales. Panel B is an example of a negative linear relationship. As X increases, we note that the values of Y are decreasing. An example of this type of relationship might be the price of a particular product and the amount of sales. Panel C shows a set of data in which there is very little or no relationship between X and Y. High and low values of Y appear at each value of X. The data in Panel D show a positive curvilinear relationship between X and Y. The values of Y are increasing as X increases, but this increase tapers off beyond certain values of X. An example of this positive curvilinear relationship might be the age and maintenance cost of a machine. As a machine gets older, the maintenance cost may rise rapidly at first but then level off beyond a certain number of years. Panel E shows a parabolic or U-shaped relationship between X and Y. As X increases, at first Y decreases; but as X continues to increase, Y not only stops decreasing but actually increases above its minimum value. An example of this type of relationship could be the number of

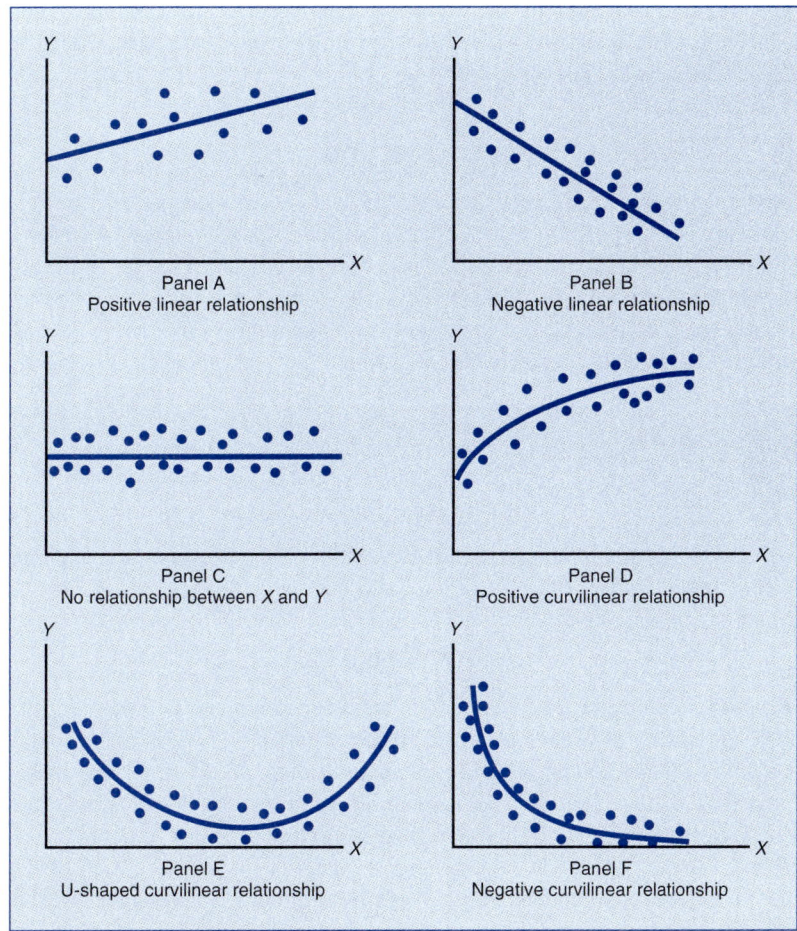

FIGURE 11.3   Examples of types of relationships found in scatter diagrams.

errors per hour at a task and the number of hours worked. The number of errors per hour would decrease as the individual becomes more proficient at the task, but then would increase beyond a certain point because of factors such as fatigue and boredom. Finally, Panel F indicates an exponential or negative curvilinear relationship between X and Y. In this case, Y decreases very rapidly as X first increases, but then decreases much less rapidly as X increases further. An example of this exponential relationship could be the resale value of a particular type of automobile and its age. In the first year, the resale value drops drastically from its original price; however, the resale value then decreases much less rapidly in subsequent years.

In this section we have briefly examined a variety of different models that could be used to represent the relationship between two variables. Although scatter diagrams can be extremely helpful in determining the mathematical form of the relationship, more sophisticated statistical procedures are available to determine the most appropriate model for a set of variables. In subsequent sections of this chapter, we shall primarily focus on building statistical models for fitting *linear* relationships between variables.

11.4 DETERMINING THE SIMPLE LINEAR REGRESSION EQUATION

If we refer to the scatter diagram in Figure 11.1 on page 535, we notice that sales appear to increase linearly as a function of the number of customers. The question that must be addressed in regression analysis involves the determination of the particular straight-line model that is the best fit to these data.

11.4.1 The Least Squares Method

In the preceding section we hypothesized a statistical model to represent the relationship between two variables in a population for a chain of package delivery stores. However, as shown in Table 11.1 on page 535, we have obtained data from only a random sample of the population. If certain assumptions are valid (see Section 11.9), the sample Y intercept (b_0) and the sample slope (b_1) can be used as estimates of the respective population parameters (β_0 and β_1). Thus, the sample regression equation representing the straight-line regression model would be

> The predicted value of Y equals the Y intercept plus the slope times the X value.
>
> $$\hat{Y}_i = b_0 + b_1 X_i \qquad (11.1a)$$

where
$$\hat{Y}_i = \text{predicted value of } Y \text{ for observation } i$$
$$X_i = \text{value of } X \text{ for observation } i$$

This equation requires the determination of two **regression coefficients**—b_0 (the Y intercept) and b_1 (the slope) in order to predict values of Y. Once b_0 and b_1 are obtained, the straight line is known and can be plotted on the scatter diagram. Then we can make a visual comparison of how well our particular statistical model (a straight line) fits the original data. That is, we can see whether the original data lie close to the fitted line or deviate greatly from the fitted line.

Simple linear regression analysis is concerned with finding the straight line that fits the data best. The best fit means that we wish to find the straight line for which the differences between the actual values (Y_i) and the values that would be predicted from the fitted line of regression ($\hat{Y}_i$) are as small as possible. Because these differences will be both positive and negative for different observations, mathematically we minimize

$$\sum_{i=1}^{n} (Y_i - \hat{Y}_i)^2$$

where
$$Y_i = \text{actual value of } Y \text{ for observation } i$$
$$\hat{Y}_i = \text{predicted value of } Y \text{ for observation } i$$

Since $\hat{Y}_i = b_0 + b_1 X_i$, we are minimizing

$$\sum_{i=1}^{n} [Y_i - (b_0 + b_1 X_i)]^2$$

which has two unknowns, b_0 and b_1.

A mathematical technique that determines the values of b_0 and b_1 that minimizes this difference is known as the **least squares method**. Any values for b_0 and b_1 other than those determined by the least squares method would result in a greater sum of squared differences between the actual value of Y and the predicted value of Y.

In using the least squares method, we obtain the following two equations, called the normal equations:

$$\sum_{i=1}^{n} Y_i = nb_0 + b_1 \sum_{i=1}^{n} X_i \tag{11.2a}$$

$$\sum_{i=1}^{n} X_i Y_i = b_0 \sum_{i=1}^{n} X_i + b_1 \sum_{i=1}^{n} X_i^2 \tag{11.2b}$$

From these two equations, we must solve for b_1 and b_0. However, in this text, we shall take the position that in solving regression equations, Excel spreadsheet software will be used to perform the (often tedious) calculations. However, to understand how the results displayed in the output of this software have been computed for the case of **simple linear regression**, we will also illustrate many of the computations involved.

Referring to Equations (11.2a) and (11.2b), since there are two equations with two unknowns, we can solve these equations simultaneously for b_1 and b_0 as follows:

$$b_1 = \frac{\sum\limits_{i=1}^{n} X_i Y_i - n\overline{X}\,\overline{Y}}{\sum\limits_{i=1}^{n} X_i^2 - n\overline{X}^2} \tag{11.3}$$

and

$$b_0 = \overline{Y} - b_1 \overline{X} \tag{11.4}$$

where
$$\bar{Y} = \frac{\sum\limits_{i=1}^{n} Y_i}{n} \quad \text{and} \quad \bar{X} = \frac{\sum\limits_{i=1}^{n} X_i}{n}$$

Examining Equations (11.3) and (11.4), we see that there are five quantities that must be calculated to determine b_1 and b_0. These are n, the sample size; $\sum\limits_{i=1}^{n} X_i$, the sum of the X values; $\sum\limits_{i=1}^{n} Y_i$, the sum of the Y values; $\sum\limits_{i=1}^{n} X_i^2$, the sum of the squared X values, and $\sum\limits_{i=1}^{n} X_i Y_i$, the sum of the cross product of X and Y. For our data in Table 11.1, the number of customers is used to predict the weekly sales in a store. The computation of the various sums needed (including $\sum\limits_{i=1}^{n} Y_i^2$, the sum of the squared Y values that will be used in Section 11.5) are presented in Table 11.2.

Using Equations (11.3) and (11.4), we can compute the values of b_0 and b_1:

$$b_1 = \frac{\sum\limits_{i=1}^{n} X_i Y_i - n\bar{X}\bar{Y}}{\sum\limits_{i=1}^{n} X_i^2 - n\bar{X}^2}$$

Table 11.2 **Computations for the package delivery sales data.**

Store	Customers X	Sales Y	X^2	Y^2	XY
1	907	11.20	822,649	125.4400	10,158.40
2	926	11.05	857,476	122.1025	10,232.30
3	506	6.84	256,036	46.7856	3,461.04
4	741	9.21	549,081	84.8241	6,824.61
5	789	9.42	622,521	88.7364	7,432.38
6	889	10.08	790,321	101.6064	8,961.12
7	874	9.45	763,876	89.3025	8,259.30
8	510	6.73	260,100	45.2929	3,432.30
9	529	7.24	279,841	52.4176	3,829.96
10	420	6.12	176,400	37.4544	2,570.40
11	679	7.63	461,041	58.2169	5,180.77
12	872	9.43	760,384	88.9249	8,222.96
13	924	9.46	853,776	89.4916	8,741.04
14	607	7.64	368,449	58.3696	4,637.48
15	452	6.92	204,304	47.8864	3,127.84
16	729	8.95	531,441	80.1025	6,524.55
17	794	9.33	630,436	87.0489	7,408.02
18	844	10.23	712,336	104.6529	8,634.12
19	1,010	11.77	1,020,100	138.5329	11,887.70
20	621	7.41	385,641	54.9081	4,601.61
Totals	14,623	176.11	11,306,209	1,602.0971	134,127.90

where

$$\bar{Y} = \frac{\sum_{i=1}^{n} Y_i}{n} = \frac{176.11}{20} = 8.8055$$

$$\bar{X} = \frac{\sum_{i=1}^{n} X_i}{n} = \frac{14,623}{20} = 731.15$$

so that

$$b_1 = \frac{134,127.90 - (20)(731.15)(8.8055)}{11,306,209 - 20(731.15)^2}$$

$$= \frac{5,365.08}{614,603} = +.00873$$

and

$$b_0 = \bar{Y} - b_1\bar{X}$$

$$= 8.8055 - (.00873)(731.15) = +2.423$$

Thus, the equation for the *best* straight line for these data is

$$\hat{Y}_i = 2.423 + .00873X_i$$

The slope b_1 was computed as $+.00873$. This means that for each increase of one unit in X, the value of Y is estimated to increase by an average of .00873 unit. This means that for each increase of one customer, the fitted model predicts that the expected weekly sales are estimated to increase by .00873 thousands of dollars or $8.73 (or we can say that for each increase of 100 customers, weekly sales are expected to increase by $873). Thus, the slope can be viewed as representing the portion of the weekly sales that are estimated to vary according to the number of customers.

The Y intercept b_0 was computed to be $+2.423$ (thousands of dollars). The Y intercept represents the average value of Y when X equals 0. Since the number of customers is unlikely to be 0, this Y intercept can be viewed as expressing the portion of the weekly sales that varies with factors other than the number of customers.

The regression model that has been fit to the data can now be used to predict the weekly sales. For example, let us say that we would like to use the fitted model to predict the average weekly sales for a store with 600 customers. We can determine the predicted value by substituting $X = 600$ into our regression equation,

$$\hat{Y}_i = 2.423 + .00873(600) = 7.661$$

Thus, the predicted average weekly sales for a store with 600 customers is 7.661 thousands of dollars or $7,661.

11.4.2 Predictions in Regression Analysis: Interpolation Versus Extrapolation

When using a regression model for prediction purposes, it is important that we consider only the relevant range of the independent variable in making our predictions. This **relevant range** encompasses all values from the smallest to the largest X used in developing the regression model. Hence, when predicting Y for a given value of X, we may *interpolate* within this relevant range of the X values, but we may not *extrapolate* beyond the range of X values. For example, when we use the number of customers to predict weekly sales, we note from Table 11.1 that the number of customers varies from 420 to 1,010. Therefore, predictions of weekly sales should be made only for stores that have between 420 and 1,010 customers. Any prediction of weekly sales outside this range of number of customers presumes that the fitted relationship holds outside the 420–1,010 range.

Problems for Section 11.4

Note: *Excel may be used to obtain the results for these problems (see Sections 11.6.3 and 11.7).*

PETFOOD.TXT

11.6 Referring to Problem 11.1 (pet food sales) on page 536
(a) Assuming a linear relationship, use the least squares method to find the regression coefficients b_0 and b_1.
(b) Interpret the meaning of the slope b_1 in this problem.
(c) Predict the average weekly sales (in hundreds of dollars) of pet food for stores with 8 feet of shelf space for pet food.
(d) Suppose that sales in store 12 are 2.6. Do parts (a)–(c) with this value and compare the results.

S-SITE.XLS

11.7 Referring to Problem 11.2 (site selection) on page 536
(a) Assuming a linear relationship, use the least squares method to find the regression coefficients b_0 and b_1.
(b) Interpret the meaning of the slope b_1 in this problem.
(c) Predict the average annual sales for a store that contains 4,000 square feet.
(d) Suppose that sales in store 7 were 2,660. Do parts (a)–(c) with this value and compare the results.

WORKHRS.TXT

11.8 Referring to Problem 11.3 (production worker-hours) on page 537
(a) Assuming a linear relationship, use the least squares method to find the regression coefficients b_0 and b_1.
(b) Interpret the meaning of the Y intercept b_0 and the slope b_1 in this problem.
(c) Predict the average number of worker-hours required for a production run with a lot size of 45.
(d) Why would it not be appropriate to predict the average number of worker-hours required for a production run with lot size of 100? Explain.
(e) Suppose that the worker-hours for the lot size of 60 were 117 and 119. Do parts (a)–(c) with these values and compare the results.

TOMYIELD.TXT

11.9 Referring to Problem 11.4 (tomato yield) on page 537
(a) Assuming a linear relationship, use the least squares method to find the regression coefficients of b_0 and b_1.
(b) Interpret the meaning of the Y intercept b_0 and the slope b_1 in this problem.
(c) Predict the average yield of tomatoes for a plot that has been given 15 pounds per 100 square feet of natural organic fertilizer.
(d) Why would it not be appropriate to predict the average yield for a plot that has been fertilized with 100 pounds per 100 square feet? Explain.
(e) Suppose the yield with 40 pounds of fertilizer was 30 and 32 pounds. Do parts (a)–(c) with these values and compare the results.
(f) What would the results of (e) lead you to think about the relationship between amount of fertilizer and yield for amounts of 40 pounds or more?

LIMO.DAT

11.10 Referring to Problem 11.5 (airport travel) on pages 537–538
(a) Assuming a linear relationship, use the least squares method to find the regression coefficients of b_0 and b_1.
(b) Interpret the meaning of the Y intercept b_0 and the slope b_1 in this problem.
(c) Use the regression model developed in (a) to predict the number of minutes to transport someone from a location that is 21 miles from the airport.
(d) Suppose the distance for the last trip was 36.7 miles and the time was 65 minutes. Do parts (a)–(c) with these values and compare the results.
(e) What would the results of (d) lead you to think about the usefulness of the regression model?

11.5 STANDARD ERROR OF THE ESTIMATE

In the preceding section we used the least squares method to develop an equation to predict the weekly sales based on the number of customers. Although the least squares method results in the line that fits the data with the minimum amount of variation, the regression equation is not a perfect predictor, unless all the observed data points fall on the regression line. Just as we do not expect all data values to be exactly equal to their arithmetic *mean*, neither can we expect all data points to fall exactly on the regression line. The regression line serves only as an approximate predictor of a Y value for a given value of X. Therefore, we need to develop a statistic that measures the variability of the actual Y values, from the predicted Y values, in the same way that we developed (see Chapter 3) a measure of the variability of each observation around its mean. The measure of variability around the line of regression (its standard deviation) is called the **standard error of the estimate**.

The variability around the line of regression is illustrated in Figure 11.4 for the package delivery sales data. We can see from Figure 11.4 that, although the predicted line of regression falls near many of the actual values of Y, there are several values above the line of regression as well as below the line of regression, so that

$$\sum_{i=1}^{n} (Y_i - \hat{Y}_i) = 0$$

The standard error of the estimate, given by the symbol S_{YX} is defined as

$$S_{YX} = \sqrt{\frac{\sum_{i=1}^{n} (Y_i - \hat{Y}_i)^2}{n - 2}} \qquad (11.5)$$

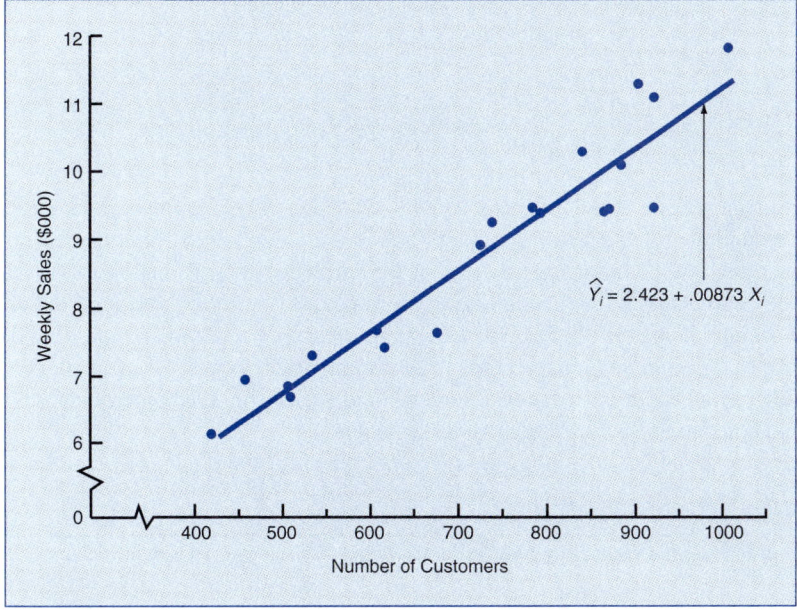

FIGURE 11.4   Scatter diagram and line of regression for the package delivery sales data.

where
$$Y_i = \text{actual value of } Y \text{ for a given } X_i$$
$$\hat{Y}_i = \text{predicted value of } Y \text{ for a given } X_i$$

The computation of the standard error of the estimate using Equation (11.5) would first require determining the predicted value of Y for each X value in the sample. The computation can be simplified because of the following identity:

$$\sum_{i=1}^{n} (Y_i - \hat{Y}_i)^2 = \sum_{i=1}^{n} Y_i^2 - b_0 \sum_{i=1}^{n} Y_i - b_1 \sum_{i=1}^{n} X_i Y_i$$

The standard error of the estimate S_{YX} can thus be obtained using the following formula:

$$S_{YX} = \sqrt{\frac{\displaystyle\sum_{i=1}^{n} Y_i^2 - b_0 \sum_{i=1}^{n} Y_i - b_1 \sum_{i=1}^{n} X_i Y_i}{n - 2}} \tag{11.6}$$

For the package delivery sales example, from Table 11.2 on page 542 we have determined that

$$\sum_{i=1}^{n} Y_i^2 = 1,602.0971 \qquad \sum_{i=1}^{n} Y_i = 176.11 \qquad \sum_{i=1}^{n} X_i Y_i = 134,127.90$$

$$b_0 = 2.423 \qquad b_1 = +.00873$$

Therefore, using Equation (11.6), the standard error of the estimate S_{YX} can be computed as

$$S_{YX} = \sqrt{\frac{\displaystyle\sum_{i=1}^{n} Y_i^2 - b_0 \sum_{i=1}^{n} Y_i - b_1 \sum_{i=1}^{n} X_i Y_i}{n - 2}}$$

$$= \sqrt{\frac{1,602.0971 - (2.423)(176.11) - (.00873)(134,127.90)}{20 - 2}}$$

$$= \sqrt{\frac{4.446}{18}} = \sqrt{.247}$$

$$= .497$$

This standard error of the estimate, equal to .497 (i.e., $497) represents a measure of the variation around the fitted line of regression. It is measured in units of the dependent variable Y. The interpretation of the standard error of the estimate, then, is analogous to that of the standard deviation. Just as the standard deviation measures variability around the arithmetic mean, the standard error of the estimate measures variability around the fitted line of regression. As we shall see in Sections 11.12–11.14, the standard error of the estimate can be used to make inferences about a predicted value of Y and to determine whether a statistically significant relationship exists between the two variables.

Problems for Section 11.5

Note: The problems in this section can be solved using Microsoft Excel (see Sections 11.6.3 and 11.7).

PETFOOD.TXT

● 11.11 Referring to the pet food sales problem (pages 536 and 544), compute the standard error of the estimate.

11.12 Referring to the site selection problem (pages 536 and 544), compute the standard error of the estimate.

11.13 Referring to the production worker-hours problem (pages 537 and 544), compute the standard error of the estimate.

11.14 Referring to the tomato yield problem (pages 537 and 544), compute the standard error of the estimate.

11.15 Referring to the airport travel problem (pages 537 and 544), compute the standard error of the estimate.

S-SITE.XLS

WORKHRS.TXT

TOMYIELD.TXT

LIMO.TXT

11.6 MEASURES OF VARIATION IN REGRESSION AND CORRELATION

11.6.1 Obtaining the Sum of Squares

To examine how well the independent variable predicts the dependent variable in our statistical model, we need to develop several measures of variation. The first measure, the **total sum of squares** (SST), is a measure of variation of the Y_i values around their mean $\bar{Y}$. In a regression analysis the total sum of squares can be subdivided into **explained variation** or **regression sum of squares** (SSR), that which is attributable to the relationship between X and Y, and **unexplained variation** or **error sum of squares** (SSE), that which is attributable to factors other than the relationship between X and Y. These different measures of variation can be seen in Figure 11.5.

The regression sum of squares (SSR) represents the difference between $\bar{Y}$ (the average value of Y) and $\hat{Y}_i$ (the value of Y that would be predicted from the regression relationship). The error sum of squares (SSE) represents that part of the variation in Y that is not explained by the regression. It is based on the difference between Y_i and $\hat{Y}_i$.

These measures of variation can be represented as follows:

$$\text{Total sum of squares} = \text{regression sum of squares} + \text{error sum of squares}$$
$$\text{SST} = \text{SSR} + \text{SSE} \tag{11.7}$$

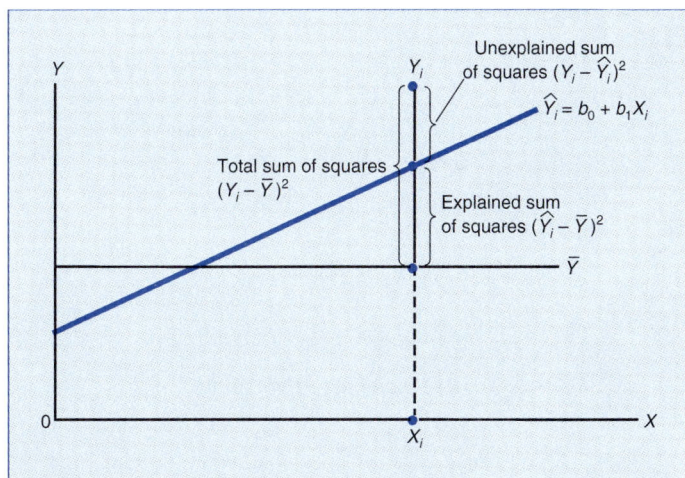

FIGURE 11.5 Measures of variation in regression.

where

$$\text{SST} = \text{total sum of squares} = \sum_{i=1}^{n}(Y_i - \overline{Y})^2 = \sum_{i=1}^{n} Y_i^2 - n\overline{Y}^2 \qquad (11.8)$$

$$\text{SSE} = \text{unexplained variation or error sum of squares}$$

$$= \sum_{i=1}^{n}(Y_i - \hat{Y}_i)^2 \qquad (11.9)$$

$$= \sum_{i=1}^{n} Y_i^2 - b_0\sum_{i=1}^{n} Y_i - b_1\sum_{i=1}^{n} X_i Y_i$$

$$\text{SSR} = \text{explained variation or regression sum of squares}$$

$$= \sum_{i=1}^{n}(\hat{Y}_i - \overline{Y})^2$$

$$= \text{SST} - \text{SSE} \qquad (11.10)$$

$$= b_0\sum_{i=1}^{n} Y_i + b_1\sum_{i=1}^{n} X_i Y_i - n\overline{Y}^2$$

Examining the unexplained variation or error sum of squares [Equation (11.9)], we may recall that $\sum_{i=1}^{n}(Y_i - \hat{Y}_i)^2$ was the numerator under the square root in the computation of the standard error of the estimate [see Equation (11.5)]. Therefore, in the process of computing the standard error of the estimate, we have already computed the following error sum of squares:

$$\text{SSE} = \sum_{i=1}^{n} Y_i^2 - b_0\sum_{i=1}^{n} Y_i - b_1\sum_{i=1}^{n} X_i Y_i$$

$$= 1{,}602.0971 - (2.423)(176.11) - (.00873)(134{,}127.90)$$

$$= 4.446$$

In addition,

$$\text{SST} = \text{total sum of squares}$$

$$= \sum_{i=1}^{n} Y_i^2 - n\overline{Y}^2$$

$$= 1{,}602.0971 - 20(8.8055)^2$$

$$= 1{,}602.0971 - 1{,}550.7366$$

$$= 51.3605$$

and

$$\text{SSR} = \text{explained variation or regression sum of squares}$$

$$= b_0 \sum_{i=1}^{n} Y_i + b_1 \sum_{i=1}^{n} X_i Y_i - n\bar{Y}^2$$

$$= (2.423)(176.11) + (.00873)(134{,}127.90) - 20(8.8055)^2$$

$$= 46.9145$$

We note also, from Equation (11.7), that

$$\text{SST} = \text{SSR} + \text{SSE}$$

$$51.3605 = 46.9145 + 4.4460$$

11.6.2 The Coefficient of Determination

Now that SSR, SSE, and SST have been defined, the **coefficient of determination** r^2 can be defined as

> The coefficient of determination is equal to the regression sum of squares divided by the total sum of squares.
>
> $$r^2 = \frac{\text{regression sum of squares}}{\text{total sum of squares}} = \frac{\text{SSR}}{\text{SST}} \qquad (11.11a)$$

Thus, the coefficient of determination measures the proportion of variation that is explained by the independent variable in the regression model. For the package delivery sales example,

$$r^2 = \frac{46.9145}{51.3605} = .913$$

Therefore, 91.3% of the variation in weekly sales can be explained by the variability in the number of customers from store to store. This is an example where there is a strong linear relationship between two variables, since the use of a regression model has reduced the variability in predicting weekly sales by 91.3%. Only 8.7% of the sample variability in weekly sales can be explained by factors other than what is accounted for by the linear regression model.

To interpret the coefficient of determination—particularly when dealing with multiple regression models—some researchers suggest that an *adjusted* r^2 be computed to reflect both the number of explanatory variables in the model and the sample size. In simple linear regresssion, we define the **adjusted** r^2 as

> $$r^2_{adj} = 1 - \left[(1 - r^2) \frac{n-1}{n-2} \right] \qquad (11.11b)$$

Thus for our package delivery sales data, since $r^2 = .913$ and $n = 20$,

$$r^2_{adj} = 1 - \left[(1 - r^2) \frac{20-1}{20-2} \right]$$

$$= 1 - \left[(1 - .913) \frac{19}{18} \right]$$

$$= 1 - .092$$

$$= .908$$

This result is similar to the one obtained without adjustment for degrees of freedom.

11.6.3 Using the Microsoft Excel Chart Wizard and TREND Function for Regression Analysis

11-6-3.XLS

In this chapter we have developed the least squares method to compute the regression coefficients and used the regression model obtained to predict the sales for a given number of customers for the package delivery stores. We can use the Chart Wizard that we previously discussed in Chapters 2 and 3 to obtain a scatter diagram and the line of regression for these data. With the PACKAGE.XLS workbook open, click on the Data sheet tab. Note that this sheet contains the data of Table 11.1 on page 535. Then, with the Data sheet still active, select Insert | Chart | As New Sheet. For the five dialog boxes of the Chart Wizard that appear, do the following:

Dialog box 1: Enter the range B1:C21 and click the Next button.

Dialog box 2: Select the *XY* (scatter) chart and click the Next button.

Dialog box 3: Select format 1 that shows unconnected points and click the Next button.

Dialog box 4: Select the Columns option button and enter 1 in the First Columns edit box and 1 in the First Rows edit box. Click the Next button.

Dialog box 5: Select No for Add a Legend and enter the Chart title Regression Analysis, Customers in the Category (*X*) edit box, and Sales in the Value (*Y*) edit box. Click the Finish button.

Figure 11.1.Excel displays the scatter diagram we have obtained. Rename the sheet as Trend.

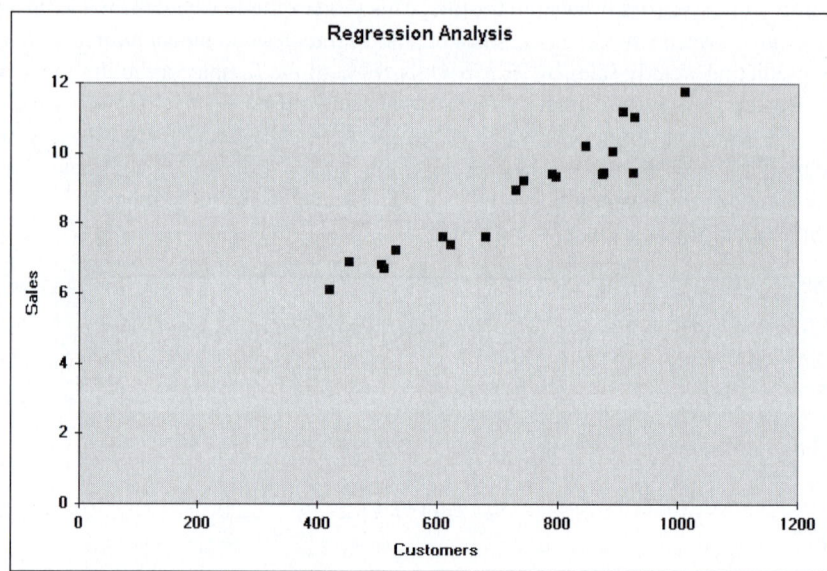

FIGURE 11.1.EXCEL Scatter diagram obtained from the Excel Chart Wizard for the package delivery sales data.

To superimpose a line of regression on this scatter diagram, select the plotted points by clicking on any single point. (Clicking causes many of the points to be highlighted.) Next, select the command Insert | Trendline. When the Trendline dialog box appears, select the Type Tab and then select Linear as the Trend/Regression Type as is illustrated in Figure 11.2.Excel. Then click the Options Tab. In the Options Tab, select the Automatic Trendline Name option button and select the Display Equation on Chart and the Display R-Squared Value on Chart check boxes (see Figure 11.3.Excel). Click the OK button. Excel adds a regression line, the regression equation, and the value of r^2 to the chart. (You can reposition the equation and r^2 by dragging them with the mouse pointer.) Figure 11.4.Excel illustrates the scatter diagram after these changes have been made.

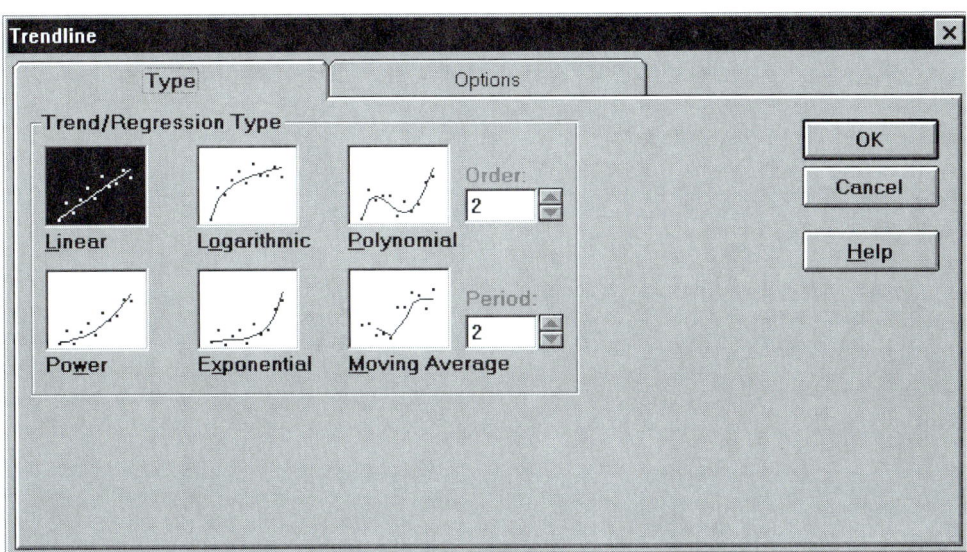

FIGURE 11.2.EXCEL   Trendline Type Tab dialog box.

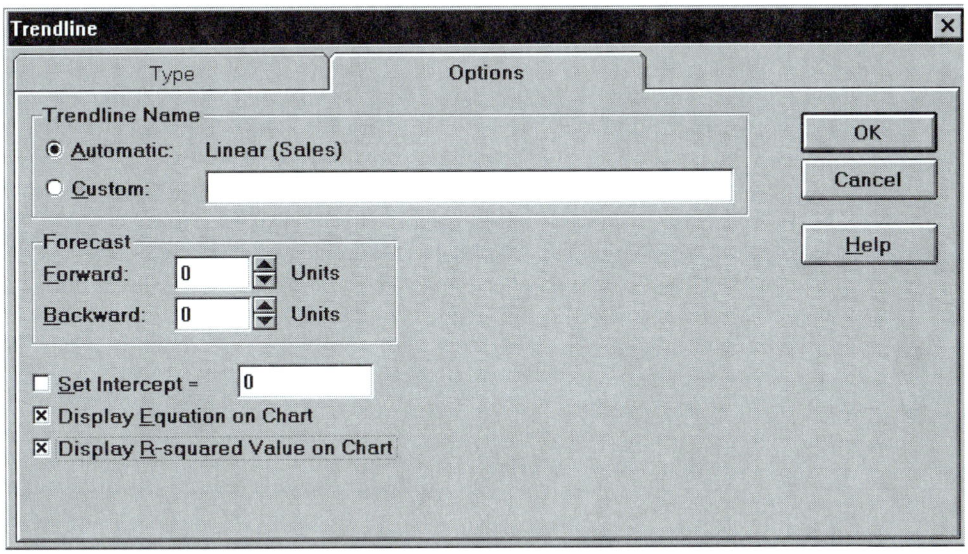

FIGURE 11.3.EXCEL   Trendline Options Tab dialog box.

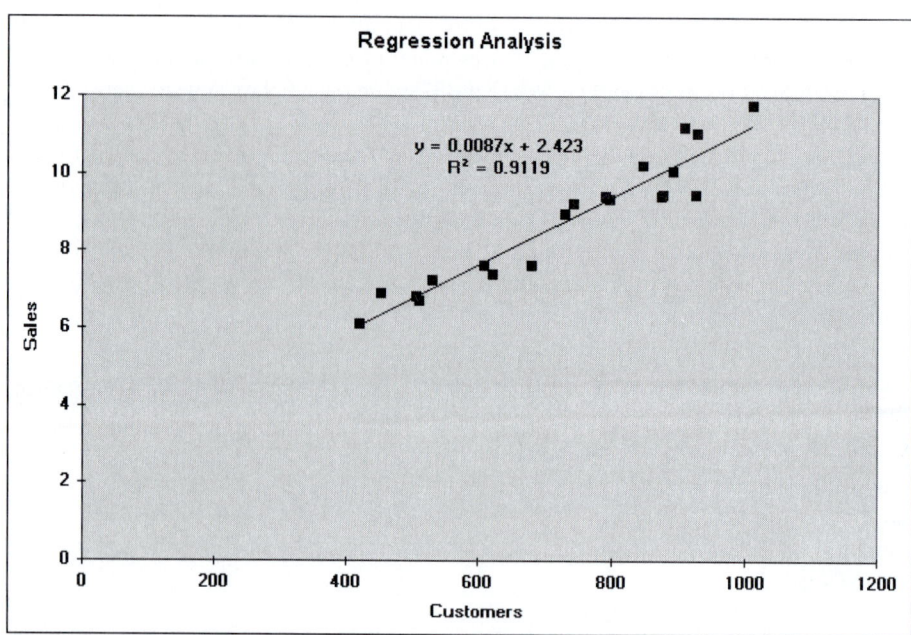

FIGURE 11.4.EXCEL   Scatter diagram with regression line and r^2 obtained from Excel's Chart Wizard for the package delivery sales data.

From Figure 11.4.Excel, we can observe that the line of regression that has been fit to these data has a Y intercept of 2.423 and a slope of 0.0087, with an R^2 of 0.9119.

▲ **WHAT IF EXAMPLE**

One of the advantages of using the Chart Wizard's TRENDLINE feature instead of the Data Analysis tool that will be discussed in Section 11.7 is that changes in the data will be immediately reflected in the fitted model. This gives us an opportunity to study the effect of changes in individual points on the fit of the regression model. Begin by removing the text of the regression equation and R^2 value by selecting them and issuing the command Edit | Clear | All. Then, for example, change the sales for the nineteenth observation to 14.77 from 11.77 on the Data sheet. The scatter diagram and fitted trend line on the Trend sheet will change as shown in Figure 11.5.Excel.

We can find the new values for the Y intercept and the regression equation by selecting the plotted points and repeating the Insert | Trendline Command explained earlier in this section. We then can observe that what seemed to be a relatively small change in a single Y value has had a major impact on the fit of the regression model. The Y intercept changed from 2.423 to 1.5779, the slope changed from .0087 to .0101, and r^2 changed from .9119 to .8054.

As before, if we were interested in seeing the effects of many different changes, we could use the Scenario Manager (see Section 1S.15) to store and use sets of alternative data values.

Now that we have fit the regression model to a set of data, we need to be able to predict the average value of Y for a given value of X as was explained in Section 11.4.1 on page 000.

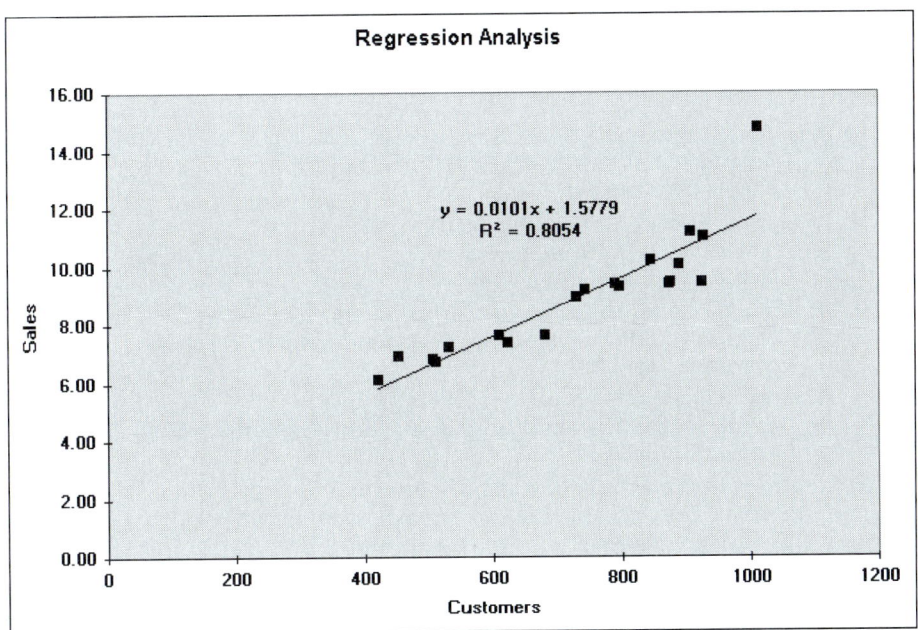

FIGURE 11.5.EXCEL Scatter diagram with regression line and r^2 obtained from Excel's Chart Wizard for the revised package delivery sales data.

This may be accomplished through the use of the Excel TREND function. The general format of the **TREND function** is

$$=\text{TREND}(\textit{range of } Y \textit{ variable, range of } X \textit{ variable, value of } X)$$

Suppose we want to use Excel to obtain a prediction of the average sales when there are 600 customers, as we did in Section 11.4.1. With the workbook developed earlier in this section still open (or with the PACKAGE.XLS workbook open), we can obtain the prediction by using the simple Calculations sheet shown in Table 11.1.Excel. To implement begin by inserting a new sheet and renaming it Calculations. Enter the number of customers in cell B3 and the formula =TREND(Data!C2:C21,Data!B2:B21,B3) in cell B4.

Figure 11.6.Excel presents the results for the data of Table 11.1.

Table 11.1.Excel Design of the Calculations sheet for the prediction of average sales.

	A	B
1	Prediction of the Average Sales	
2		
3	No. of Customers	xxx
4	Predicted Sales	=TREND(Data!C2:C21,Data!B2:B21,B3)

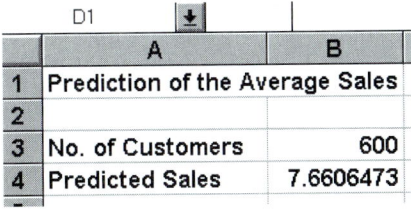

FIGURE 11.6.EXCEL Predicted value of *Y* using the TREND function for the package delivery data.

Problems for Section 11.6

Note: *The problems in this section can be solved using Microsoft Excel (see Sections 11.6.3 and 11.7).*

PETFOOD.TXT

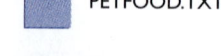

S-SITE.XLS

WORKHRS.TXT

TOMYIELD.TXT

LIMO.TXT

● 11.16 Referring to Problem 11.6 (pet food sales) on page 544
(a) Compute the coefficient of determination r^2 and interpret its meaning.
(b) Compute the adjusted r^2.

11.17 Referring to Problem 11.7 (site selection) on page 544
(a) Compute the coefficient of determination r^2 and interpret its meaning.
(b) Compute the adjusted r^2.

● 11.18 Referring to Problem 11.8 (production worker-hours) on page 544
(a) Compute the coefficient of determination r^2 and interpret its meaning.
(b) Compute the adjusted r^2.

11.19 Referring to Problem 11.9 (tomato yield) on page 544
(a) Compute the coefficient of determination r^2 and interpret its meaning.
(b) Compute the adjusted r^2.

11.20 Referring to Problem 11.10 (airport travel) on page 544
(a) Compute the coefficient of determination r^2 and interpret its meaning.
(b) Compute the adjusted r^2.

11.21 When will the unexplained variation or error sum of squares be equal to 0?

11.22 When will the explained variation or sum of squares due to regression be equal to 0?

11.7 USING THE DATA ANALYSIS TOOL FOR REGRESSION

11-7.XLS

The Data Analysis Tool of Microsoft Excel can be used instead of the TREND function to obtain a more complete regression analysis. To access the **Regression option** of Data Analysis, open the PACKAGE.XLS workbook, containing the package delivery data of Table 11.1 on page 535 and click the Data sheet tab. Select Tools | Data Analysis, select Regression in the Analysis Tools list box, and click the OK button. In the Regression dialog box that appears, do the following:

(a) Enter C1:C21 in the Input Y Range edit box.

(b) Enter B1:B21 in the Input X Range edit box.

(c) Select the Labels check box.

(d) Select the Confidence Level check box and set the level to 95%. (This will provide a 95% confidence interval estimate for the regression coefficients, which will be studied in Section 11.14.)

(e) Select the New Worksheet Ply option button and enter Regression as the name.

(f) Select the Residuals, Standardized Residuals, and Residual Plots check boxes. This information will be studied in Section 11.10. Select the Line Fit Plots check box if you wish to obtain a scatter diagram with a fitted regression line, as was illustrated in Figure 11.4.Excel on page 000. The dialog box should now appear similar to the one illustrated in Figure 11.7.Excel. Click the OK button to have Excel perform the regression analysis.

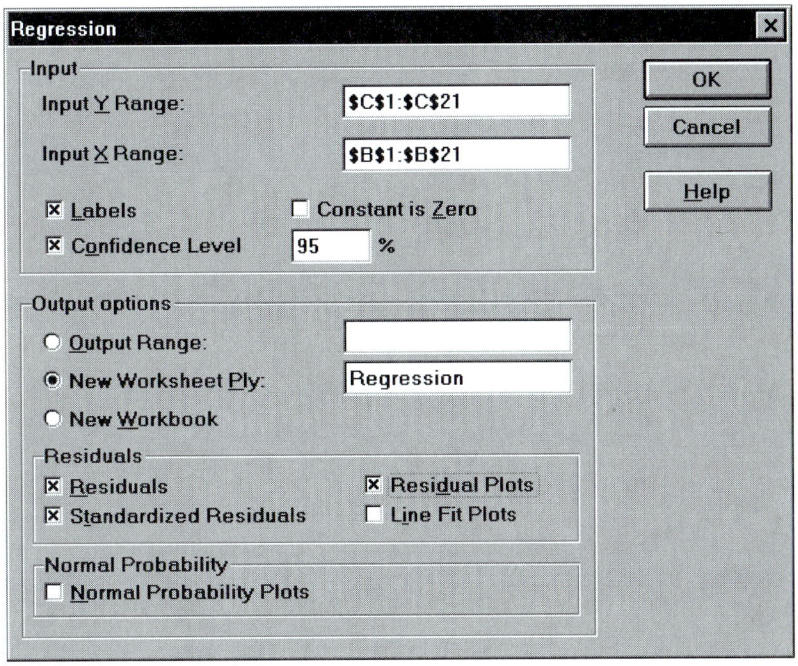

FIGURE 11.7.EXCEL   Regression dialog box from Excel's Data Analysis tool.

Figure 11.8.Excel presents the results. Note that the Residuals plot will appear on the right-hand side of the sheet. (You may need to use the horizontal scroll bars in order to see the plot on the screen.)

The text portion of the Panel A output is divided into four areas. The first area, called regression statistics, provides the values of the coefficient of correlation r, the coefficient of determination r^2, the adjusted r^2, the standard error of the estimate (labeled Standard Error), and the sample size (labeled Observations). The second area presents output in a table called ANOVA, which will be discussed in detail in Chapter 12 when we cover multiple regression. However, if we examine the column labeled SS (for sum of squares), we can find the regression sum of squares (SSR) equal to 46.8335409, the residual or error sum of squares (SSE) equal to 4.5269541, and the total sum of squares (SST) equal to 51.360495. Any minor differences compared to those obtained in Section 11.6 are due to the fact that when we use calculator formulas, rounding errors can occur because the number of significant digits that can be used is limited.

The third area of the output provides information concerning the regression coefficients b_0 and b_1. The column labeled coefficients provides the values of the Y intercept b_0 and the slope b_1. The Y intercept is equal to 2.4230444, and the slope is equal to 0.00872934. Once again, any minor differences from those obtained in Section 11.4 are due to rounding errors. The remainder of the information in this portion of the output will be discussed in Section 11.14.

The fourth area of Panel A includes the predicted sales for each of the 20 data points along with values for the residuals and standardized residuals that will be discussed in Section 11.10. Panel B, containing a plot of the residuals and the X values, will also be discussed in Section 11.10.

	A	B	C	D	E	F
1	Summary Output					
2						
3	Regression Statistics					
4	Multiple R	0.9549132				
5	R Square	0.91185922				
6	Adjusted R Square	0.90696251				
7	Standard Error SYX	0.501495215				
8	Observations	20				
9						
10	ANOVA					
11		df	SS	MS	F	Significance F
12	Regression	1	SSR 46.8335409	46.8335409	186.22	6.20621E-11
13	Residual	18	SSE 4.526954104	0.25149745		
14	Total	19	SST 51.360495			
15						
16		Coefficients	Standard Error	t Stat	P-value	
17	Intercept b_0	2.423044396	0.480964609	5.037885009	8.5539E-05	
18	Customers b_1	0.008729338	0.00063969	13.64619912	6.2062E-11	
19						
20						
21						
22	Residual Output					
23						
24	Observation	Predicted Sales	Residuals	Standard Residuals		
25	1	10.34055412	0.859445883	1.713766867		
26	2	10.50641154	0.543588457	1.083935482		
27	3	6.840089511	-8.95107E-05	-0.000178488		
28	4	8.891483981	0.318516019	0.635132719		
29	5	9.310492213	0.109507787	0.218362576		
30	6	10.18342603	-0.10342603	-0.206235329		
31	7	10.05248596	-0.602485958	-1.201379276		
32	8	6.875006863	-0.145006863	-0.289149047		
33	9	7.040864289	0.199135711	0.397083971		
34	10	6.089366428	0.030633572	0.061084475		
35	11	8.350265014	-0.720265014	-1.43623507		
36	12	10.03502728	-0.605027281	-1.206446769		
37	13	10.48895287	-1.028952866	-2.05177006		
38	14	7.721752666	-0.081752666	-0.163017839		
39	15	6.368705249	0.55129475I	1.099302116		
40	16	8.786731923	0.163268077	0.325562582		
41	17	9.354138904	-0.024138904	-0.048133867		
42	18	9.790605813	0.439394187	0.876168256		
43	19	11.23967595	0.530324051	1.057485766		
44	20	7.8439634	-0.4339634	-0.865339066		

FIGURE 11.8.EXCEL   Regression analysis output obtained from Excel's Data Analysis tool for the package delivery sales data. Panel A.

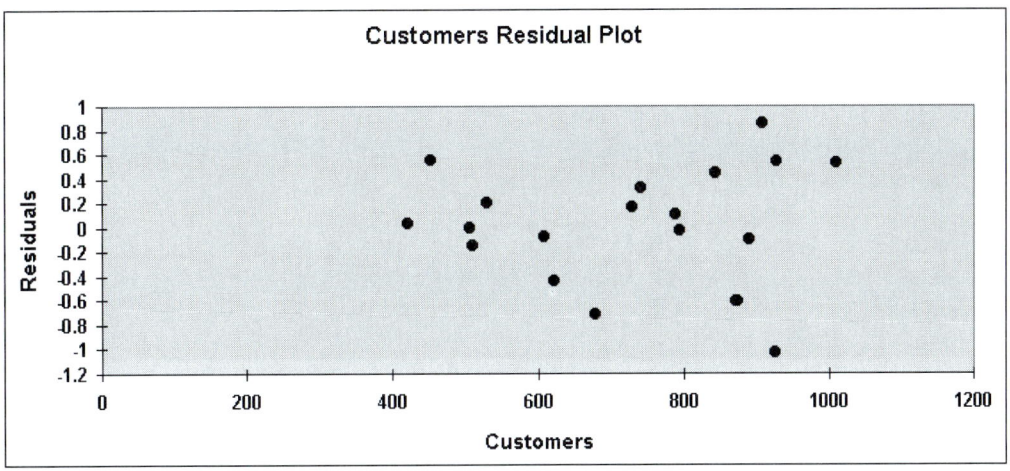

FIGURE 11.8.EXCEL   Panel B.

CORRELATION—MEASURING THE STRENGTH OF THE ASSOCIATION

11.8.1   The Correlation Coefficient

In our discussion of the relationship between two variables thus far, we have been concerned with the prediction of the dependent variable Y based on the independent variable X. In contrast to a regression analysis, in a correlation analysis we are only interested in measuring the degree of association between two variables.

The strength of a relationship between two variables in a population is usually measured by the **coefficient of correlation** ρ, whose values range from -1 for perfect negative correlation up to $+1$ for perfect positive correlation. Figure 11.6 illustrates these three different types of association between variables. In Panel A of Figure 11.6, there is a perfect negative linear relationship between X and Y so that Y will decrease in a perfectly predictable manner as X

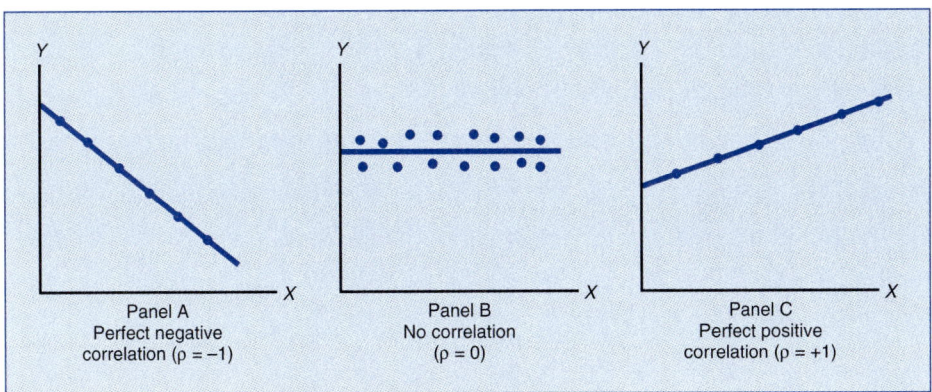

FIGURE 11.6   **Types of association between variables.**

increases. Panel B is an example in which there is no relationship between X and Y. As X increases, there is no change in Y, so there is no association between the values of X and the values of Y. Panel C depicts a perfect positive correlation between X and Y. In this case, Y increases in a perfectly predictable manner as X increases.

For regression-oriented problems, the sample coefficient of correlation (r) may be obtained from Equation (11.11a) as follows:

$$r^2 = \frac{\text{regression sum of squares}}{\text{total sum of squares}} = \frac{\text{SSR}}{\text{SST}}$$

so that

$$r = \sqrt{r^2} \tag{11.12}$$

In simple linear regression, r takes the sign of b_1. If b_1 is positive, r is positive. If b_1 is negative, r is negative. If b_1 is 0, r is 0.

In the package delivery sales example, since $r^2 = .913$ and the slope b_1 is positive, the coefficient of correlation is computed as +.956. The closeness of the correlation coefficient to +1.0 implies a strong association between number of customers and weekly sales.

We have now computed and interpreted the correlation coefficient in terms of its regression viewpoint. As we mentioned at the beginning of this chapter, however, regression and correlation are two separate techniques, with regression being concerned with prediction and correlation with association. In many applications, we are concerned only with measuring association between variables, not with using one variable to predict another.

If only a correlation analysis is being performed on a set of data, the sample correlation coefficient r can be computed directly using the following formula,

$$r = \frac{\displaystyle\sum_{i=1}^{n}(X_i - \bar{X})(Y_i - \bar{Y})}{\sqrt{\displaystyle\sum_{i=1}^{n}(X_i - \bar{X})^2}\sqrt{\displaystyle\sum_{i=1}^{n}(Y_i - \bar{Y})^2}} \tag{11.13a}$$

or, alternatively, using the "calculator" formula:

$$r = \frac{\displaystyle\sum_{i=1}^{n}X_i Y_i - n\bar{X}\bar{Y}}{\sqrt{\displaystyle\sum_{i=1}^{n}X_i^2 - n\bar{X}^2}\sqrt{\displaystyle\sum_{i=1}^{n}Y_i^2 - n\bar{Y}^2}} \tag{11.13b}$$

To illustrate such an example, suppose we want to measure the strength of the association in the price of two different grocery items in various cities throughout the world. The price of a six-pack of a brand-name cola soft drink and of 1 pound of chicken is determined at a supermarket located in a sample of nine different cities. The results are summarized in Table 11.3.

Table 11.3 Price (in $) of a six-pack of a brand-name cola soft drink and of 1 pound of chicken in a sample of nine cities.

City	Brand-Name Cola (X) (1 six-pack)	Chicken (Y) (1 pound)
Frankfurt	3.27	3.06
Hong Kong	2.22	2.34
London	2.28	2.27
Manila	3.04	1.51
Mexico City	2.33	1.87
New York	2.69	1.65
Paris	4.07	3.09
Sydney	2.78	2.36
Tokyo	5.97	4.85

For the data of Table 11.3, we compute the following values:

$$\sum_{i=1}^{n} X_i = 28.65 \qquad \sum_{i=1}^{n} X_i^2 = 102.66 \qquad \sum_{i=1}^{n} Y_i = 23.00$$

$$n = 9 \qquad \sum_{i=1}^{n} Y_i^2 = 67.132 \qquad \sum_{i=1}^{n} X_i Y_i = 81.854$$

From this, we obtain

$$\overline{X} = \frac{28.65}{9} = 3.183$$

$$\overline{Y} = \frac{23.00}{9} = 2.5556$$

so that from Equation (11.13b)

$$r = \frac{\sum_{i=1}^{n} X_i Y_i - n\overline{X}\,\overline{Y}}{\sqrt{\sum_{i=1}^{n} X_i^2 - n\overline{X}^2}\,\sqrt{\sum_{i=1}^{n} Y_i^2 - n\overline{Y}^2}}$$

$$= \frac{81.854 - 9(3.183)(2.5556)}{\sqrt{102.66 - 9(3.183)^2}\,\sqrt{67.132 - 9(2.5556)^2}}$$

$$= \frac{81.8540 - 73.2172}{\sqrt{11.4594}\,\sqrt{8.3522}}$$

$$= +.883$$

The coefficient of correlation, $r = +.883$, between the price of a brand-name cola soft drink and of chicken indicates a very strong association. A higher price of the cola is strongly associated with a higher price of chicken. In Section 11.14, we will use these sample results to determine whether there is any evidence of a significant association between these variables in the population.

11.8.2 Using the Microsoft Excel CORREL Function for Correlation Analysis

The most direct way to obtain the correlation coefficient between two variables with Excel is to use the CORREL function. The general format of the CORREL function is

$$=CORREL(\text{range of Y variable, range of X variable})$$

11-8-2.XLS

The 11-8-2.XLS workbook illustrates the use of this function to obtain the correlation coefficient for the international grocery price data of Table 11.3 on page 000. Open this workbook and click the Calculations sheet tab. Note for this small example the data of Table 11.3 have been placed on the Calculations sheet and not on a separate Data sheet as they could have been. Cell B14 computes the correlation coefficient using the formula =CORREL(B4:B12,C4:C12) because the range of the Y variable is B4:B12 and the range of the X variable is C4:C12. Figure 11.9.Excel presents the correlation coefficient obtained.

	A	B	C
1	Correlation of Cola and Chicken		
2			
3	City	Cola	Chicken
4	Frankfurt	3.27	3.06
5	Hong Kong	2.22	2.34
6	London	2.28	2.27
7	Manila	3.04	1.51
8	Mexico City	2.33	1.87
9	New York	2.69	1.65
10	Paris	4.07	3.09
11	Sydney	2.78	2.36
12	Tokyo	5.97	4.85
13			
14	Correlation	0.8828512	

FIGURE 11.9.EXCEL Correlation coefficient obtained from Excel's **CORREL** function for the international grocery price data.

Problems for Section 11.8

Note: The problems in this section can be solved using Microsoft Excel.

11.23 Under what circumstances will the coefficient of correlation be negative?

● 11.24 Referring to Problem 11.16 (pet food sales) on page 554, compute the coefficient of correlation.

11.25 Referring to Problem 11.17 (site selection) on page 554, compute the coefficient of correlation.

● 11.26 Referring to Problem 11.18 (production worker-hours) on page 554, compute the coefficient of correlation.

11.27 Referring to Problem 11.19 (tomato yield) on page 554, compute the coefficient of correlation.

11.28 Referring to Problem 11.20 (airport travel) on page 554, compute the coefficient of correlation.

11.29 Suppose we also want to measure the strength of the association in the price (in dollars) of a six-pack of a brand-name cola soft drink and 100 tablets of a brand-name pain reliever worldwide. We again determine these prices in a nine-city sample of local supermarkets. The results are as follows:

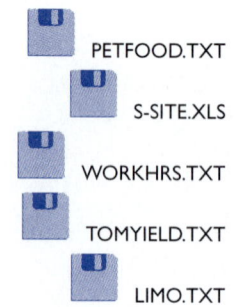

PETFOOD.TXT

S-SITE.XLS

WORKHRS.TXT

TOMYIELD.TXT

LIMO.TXT

City	Brand-Name Cola (six-pack)	Pain Reliever (100 tablets)
Frankfurt	3.27	17.22
Hong Kong	2.22	6.21
London	2.28	9.17
Manila	3.04	14.61
Mexico City	2.33	4.85
New York	2.69	6.09
Paris	4.07	13.08
Sydney	2.78	8.04
Tokyo	5.97	8.39

(a) Compute the coefficient of correlation r between the price of the cola soft drink and the pain reliever.

(b) Is the price of the cola soft drink more correlated with the price of chicken or with that of the pain reliever? Explain.

11.30 Suppose we also want to measure the strength of the association in the price (in $) of a women's haircut and a men's brand-name dress shirt in a sample of nine different international cities. The results are as follows:

City	Women's Haircut	Men's Dress Shirt
Frankfurt	29.85	49.41
Hong Kong	22.56	29.32
London	33.79	42.12
Manila	12.04	35.22
Mexico City	15.49	25.04
New York	34.87	37.85
Paris	27.73	55.28
Sydney	25.64	38.58
Tokyo	27.45	38.69

Compute the coefficient of correlation r between the women's haircut and the men's dress shirt.

11.9 ASSUMPTIONS OF REGRESSION AND CORRELATION

In our investigations into hypothesis testing and the analysis of variance, we have noted that the appropriate application of a particular statistical procedure is dependent on how well a set of assumptions for that procedure are met. The assumptions necessary for regression and correlation analysis are analogous to those of the analysis of variance, since they fall under the general heading of linear models (Reference 12). Although there are some differences in the assumptions made by the regression model and by correlation (see Reference 12), this topic is beyond the scope of this text and we will consider only the former.

The four major **assumptions of regression** are

1. Normality

2. Homoscedasticity

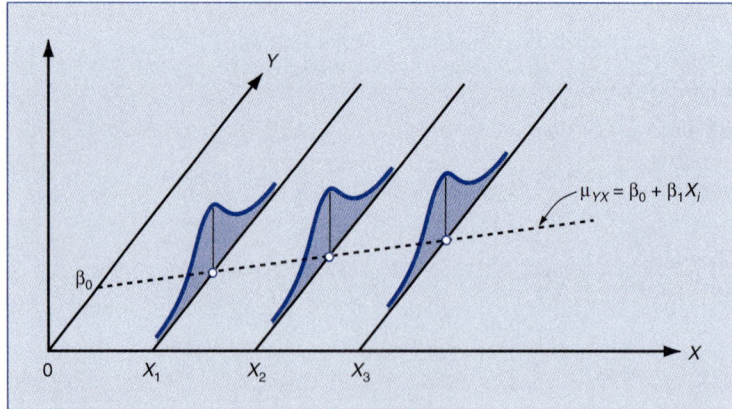

FIGURE 11.7   Assumptions of regression.

3. Independence of errors

4. Linearity

The first assumption, **normality**, requires that the values of Y be normally distributed at each value of X (see Figure 11.7). Like the t test and the ANOVA F test, regression analysis is fairly robust against departures from the normality assumption. As long as the distribution of Y_i values around each level of X is not extremely different from a normal distribution, inferences about the line of regression and the regression coefficients will not be seriously affected.

The second assumption, **homoscedasticity**, requires that the variation around the line of regression be constant for all values of X. This means that Y varies the same amount when X is a low value as when X is a high value (see Figure 11.7). The homoscedasticity assumption is important for using the least squares method of determining the regression coefficients. If there are serious departures from this assumption, either data transformations or weighted least squares methods (Reference 12) can be applied.

The third assumption, **independence of errors**, requires that the error (residual difference between observed and predicted values of Y) should be independent for each value of X. This assumption often refers to data that are collected over a period of time. When data are collected in this manner, the residuals for a particular time period are often correlated with those of the previous time period.

The fourth assumption, **linearity**, states that the relationship among variables is linear. Two variables could be perfectly related in a nonlinear fashion, and the linear correlation coefficient would be 0, indicating no relationship. Such nonlinear models will be discussed in Sections 12.10 and 12.12.

11.10   RESIDUAL ANALYSIS

11.10.1   Introduction

In the preceding discussion of the package delivery sales data, we have relied on a simple regression model in which the dependent variable is predicted based on a straight-line relationship with a single independent variable. In this section we shall use a graphical approach called **residual analysis** to evaluate the appropriateness of the regression model that has been fitted to the data. In addition, this approach will also allow us to study potential violations in the assumptions of our regression model (see Section 11.9).

11.10.2   Evaluating the Aptness of the Fitted Model

The **residual** or estimated error values (e_i) are defined as the difference between the observed (Y_i) and predicted ($\hat{Y}_i$) values of the dependent variable for given values X_i. Thus, the following definition applies:

> The residual equals the observed value of Y minus the predicted value of Y.
>
> $$e_i = Y_i - \hat{Y}_i \qquad (11.14)$$

We may evaluate the aptness of the fitted regression model by plotting the residuals on the vertical axis against the corresponding X_i values of the independent variable on the horizontal axis. If the fitted model is appropriate for the data, *there will be no apparent pattern in this plot of the residuals versus X_i*. However, if the fitted model is not appropriate, there will be a relationship between the X_i values and the residuals e_i. Such a pattern can be observed in Figure 11.8. Figure 11.8(a) depicts a situation in which there is a significant simple linear relationship between X and Y. However, a curvilinear model between the two variables seems more appropriate. This effect is highlighted in Figure 11.8(b), the residual plot of e_i versus X_i. In (b) there is a clear curvilinear effect between X_i and e_i. By plotting the residuals, we have essentially filtered out or removed the *linear* trend of X with Y, thereby exposing the lack of fit in the simple linear model. Thus, from (a) and (b), we can conclude that the curvilinear model is a better fit and should be evaluated in place of the simple linear model (see Section 12.10 for further discussion of fitting curvilinear models).

Having considered Figure 11.8, let us return to the evaluation of the package delivery sales data. Table 11.4 on page 564 lists the observed, predicted, and residual values of the response variable (weekly sales) in the simple linear model we have fitted. In addition to the residuals, we can also compute the standardized residuals and the Studentized residuals. The **standardized residuals** represent each residual divided by its standard error. The **Studentized residuals**, expressed as Equation (11.15), are the standardized residuals adjusted for the distance from the average X value.

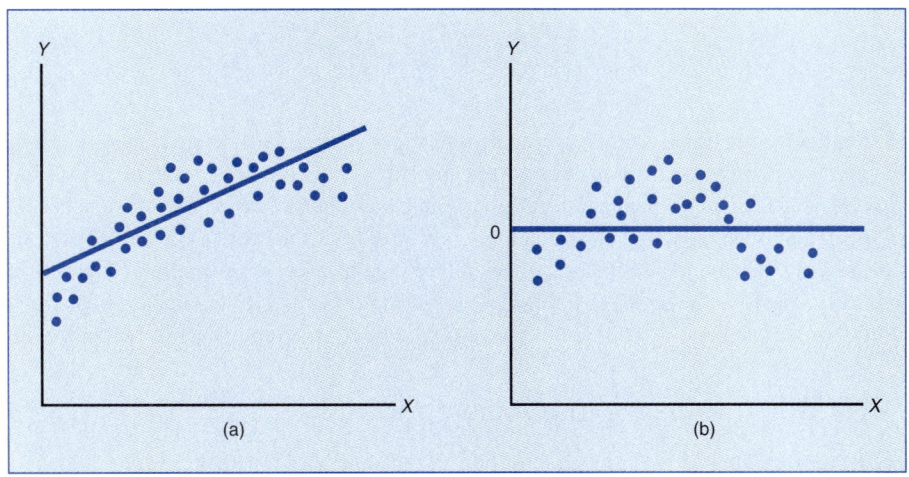

FIGURE 11.8   **Studying the appropriateness of the simple linear regression model.**

Table 11.4   Observed, predicted, and residual values for the package delivery sales data.

Observation	Customers X_i	Weekly Sales Observed	Weekly Sales Predicted	Weekly Sales Residual	Studentized Residual, SR_i
1	907	11.200	10.341	0.859	1.81
2	926	11.050	10.506	0.544	1.15
3	506	6.840	6.840	−0.000	−0.00
4	741	9.210	8.891	0.319	0.65
5	789	9.420	9.310	0.110	0.22
6	889	10.080	10.183	−0.103	−0.22
7	874	9.450	10.052	−0.602	−1.25
8	510	6.730	6.875	−0.145	−0.31
9	529	7.240	7.041	0.199	0.42
10	420	6.120	6.089	0.031	0.07
11	679	7.630	8.350	−0.720	−1.48
12	872	9.430	10.035	−0.605	−1.26
13	924	9.460	10.489	−1.029	−2.18
14	607	7.640	7.722	−0.082	−0.17
15	452	6.920	6.369	0.551	1.21
16	729	8.950	8.787	0.163	0.33
17	794	9.330	9.354	−0.024	−0.05
18	844	10.230	9.791	0.439	0.91
19	1,010	11.770	11.240	0.530	1.17
20	621	7.410	7.844	−0.434	−0.90

$$\text{Studentized Residual} = SR_i = \frac{e_i}{S_{YX}\sqrt{1 - h_i}} \qquad (11.15)$$

where

$$h_i = \frac{1}{n} + \frac{(X_i - \bar{X})^2}{\sum_{i=1}^{n}(X_i - \bar{X})^2}$$

These Studentized residuals allow us to consider the magnitude of the residuals in units that reflect the standardized variation around the line of regression. The Studentized residuals have been plotted against the independent variable (number of customers) in Figure 11.9. From this we may observe that although there is widespread scatter in the residual plot, there is no apparent pattern or relationship between the Studentized residuals and X_i. The residuals appear to be evenly spread above and below 0 for the differing values of X. Thus, we may conclude for the package delivery sales data that the fitted model appears to be appropriate.

11.10.3   Evaluating the Assumptions

● **Homoscedasticity**   The assumption of homoscedasticity (see Section 11.9) can also be evaluated from a plot of SR_i with X_i. For the package delivery sales data, there do not appear to be major differences in the variability of SR_i for different X_i values as is the case in

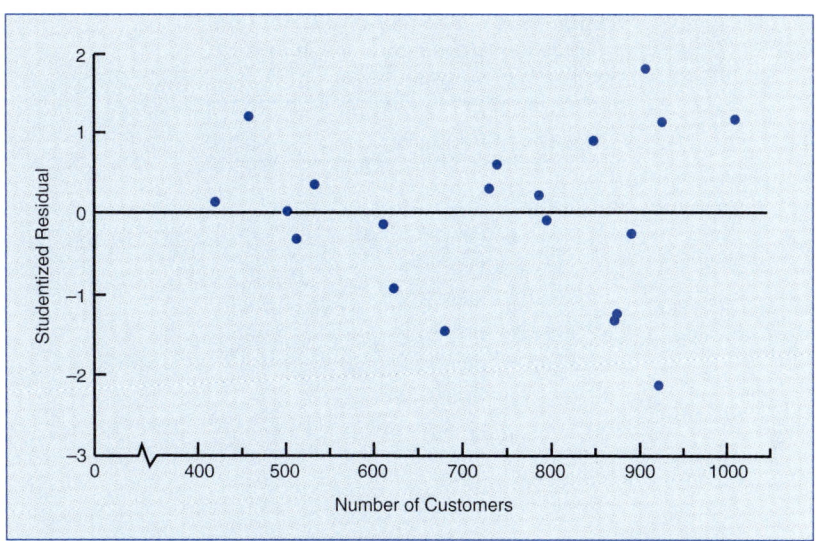

FIGURE 11.9 Plotting the Studentized residuals versus the number of customers in the package delivery sales example.

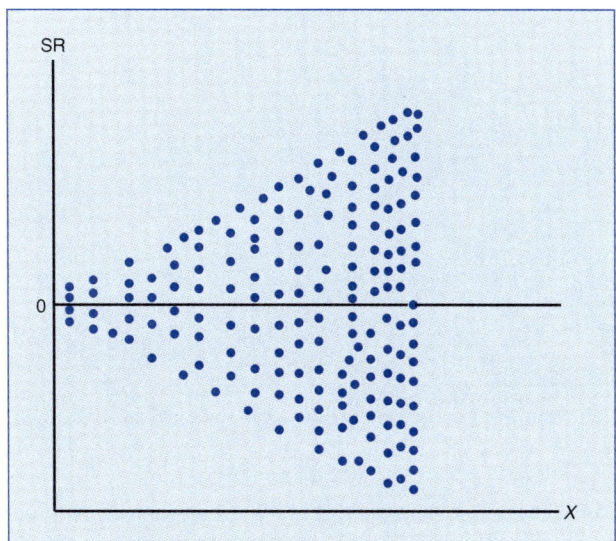

FIGURE 11.10 Violations in homoscedasticity.

Figure 11.9. Thus, we may conclude that for our fitted model there is no apparent violation in the assumption of equal variance at each level of X.

If we wish to observe a case in which the homoscedasticity assumption is violated, we should examine the *hypothetical* plot of SR_i with X_i in Figure 11.10. In this hypothetical plot, there appears to be a *fanning effect* in which the variability of the residuals increases as X increases, demonstrating the lack of homogeneity in the variances of Y_i at each level of X.

● **Normality** The normality assumption of regression (see Section 11.9) can also be evaluated from a residual analysis by tallying the Studentized residuals into a frequency distribution and displaying the results in a histogram (see Chapter 2).

For the package delivery sales data, the Studentized residuals have been tallied into a frequency distribution as indicated in Table 11.5 with the results displayed in Figure 11.11. It is

Table 11.5 Frequency distribution of 20 Studentized residual values for the package delivery data.

Studentized Residuals	Number
−2.8 but less than −2.0	1
−2.0 but less than −1.2	3
−1.2 but less than −0.4	2
−0.4 but less than +0.4	8
+0.4 but less than +1.2	4
+1.2 but less than +2.0	2
+2.0 but less than +2.8	0
Totals	20

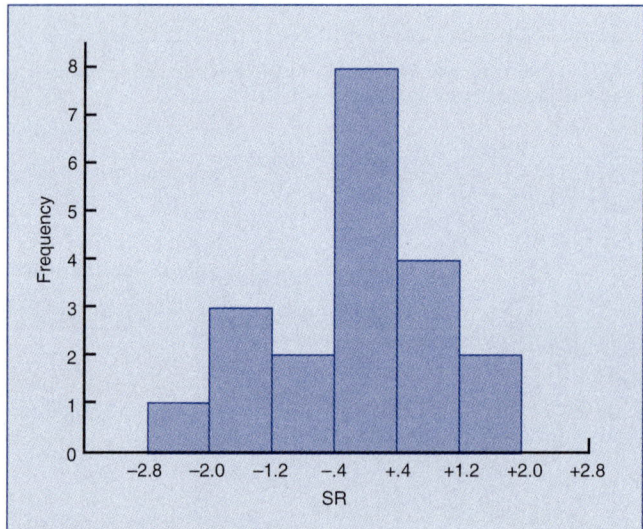

FIGURE 11.11 Plotting the Studentized residuals for the package delivery data.

difficult to evaluate the normality assumption for a sample of only 20 observations, and formal test procedures are beyond the scope of this text (see Reference 13). Although we could have also developed a normal probability plot (see Section 5.6), we can see from Figure 11.11 that the data appear to be approximately bell shaped. Thus, it seems reasonable to conclude that there is no overwhelming evidence of a violation of the normality assumption.

● **Independence** The independence assumption discussed in Section 11.9 can be evaluated by plotting the residuals in the order or sequence in which the observed data were obtained. Data collected over periods of time often exhibit an *autocorrelation* effect among successive observations. That is, there exists a correlation between a particular observation and the values that precede and succeed it. Such patterns, which violate the assumption of independence, are readily apparent in the plot of the residuals versus the time in which they were collected. This effect may be measured by the Durbin-Watson statistic, which will be the subject of Section 11.11.

11.10.4   Using Microsoft Excel for Residual Analysis

Now that we have used residual analysis to evaluate the aptness of the regression model and to test the assumptions of the model, we can determine how Excel can be accessed to obtain the pertinent charts. In Section 11.7, we used the regression option of the Data Analysis tool to obtain the output illustrated in Figure 11.8.Excel on page 556.

The residual plot provided by Excel is illustrated in Panel B of Figure 11.8.Excel. The residuals are plotted on the vertical axis, and the X variable, number of customers, is plotted on the X-axis. Figure 11.8.Excel is similar to Figure 11.9 on page 565 except that the residuals have been used in place of the Studentized residuals. We can also observe from Panel A that the residuals and standardized residuals for each observation are provided in the output.

Problems for Section 11.10

Note: *The problems in this section can be solved using Microsoft Excel.*

- **11.31** Referring to the pet food sales problem (pages 536, 544, and 546), perform a residual analysis on your results and determine the adequacy of the fit of the model.
- **11.32** Referring to the site selection problem (pages 536, 544, and 547), perform a residual analysis on your results and determine the adequacy of the fit of the model.
- **11.33** Referring to the production worker-hours problem (pages 537, 544, and 547), perform a residual analysis on your results and determine the adequacy of the fit of the model.
- **11.34** Referring to the tomato yield problem (pages 537, 544, and 547), perform a residual analysis on your results and determine the adequacy of the fit of the model.
- **11.35** Referring to the airport travel problem (pages 537, 544, and 547), perform a residual analysis on your results and determine the adequacy of the fit of the model.

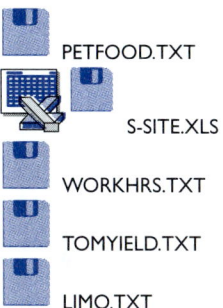

PETFOOD.TXT

S-SITE.XLS

WORKHRS.TXT

TOMYIELD.TXT

LIMO.TXT

11.11   MEASURING AUTOCORRELATION: THE DURBIN-WATSON STATISTIC

11.11.1   Introduction

One of the assumptions of the basic regression model we have been considering is the independence of the residuals. This assumption is often violated when data are collected over sequential periods of time, because a residual at any one point in time may tend to be similar to residuals at adjacent points in time. Thus, positive residuals would be more likely followed by positive residuals, and negative residuals would be more likely followed by negative residuals. Such a pattern in the residuals is called **autocorrelation**. When substantial autocorrelation is present in a set of data, the validity of a fitted regression model may be in serious doubt.

11.11.2   Residual Plots to Detect Autocorrelation

As mentioned in Section 11.10, the easiest way to detect autocorrelation in a set of data is to plot the residuals or standardized residuals in time order. If a positive autocorrelation effect is present, clusters of residuals with the same sign will be present, and an apparent pattern will be readily detected. To illustrate the autocorrelation effect, we shall consider the following example.

Recall that in Sections 11.2–11.4, we developed a regression model to predict weekly sales based on the number of customers for a sample of 20 package delivery stores. Suppose that the manager of the seventeenth package delivery store listed in Table 11.1 on page 535 wants to predict weekly sales based on the number of customers for a period of 15 weeks. In this situation, since data are collected over a period of 15 *consecutive* weeks at the *same* store, we would need to be concerned with the autocorrelation effect of the residuals. The data for this store are summarized in Table 11.6. Figure 11.12 represents partial Excel output.

	A	B	C	D	E	F
1	Summary Output					
2						
3	Regression Statistics					
4	Multiple R	0.810829997				
5	R Square	0.657445284				
6	Adjusted R Square	0.631094922				
7	Standard Error	0.936036681				
8	Observations	15				
9						
10	ANOVA					
11		df	SS	MS	F	Significance F
12	Regression	1	21.86043264	21.86043264	24.95014	0.000245105
13	Residual	13	11.39014069	0.876164669		
14	Total	14	33.25057333			
15						
16		Coefficients	Standard Error	t Stat	P-value	
17	Intercept	-16.0321936	5.310167093	-3.019150493	0.009869	
18	Customers	0.030760228	0.006158189	4.995011683	0.000245	
19						
20						
21						
22	Residual Output					
23						
24	Observation	Predicted Sales	Residuals	Standard Residuals		
25	1	8.39142731	0.93857269	1.002709305		
26	2	8.545228449	-0.285228449	-0.304719307		
27	3	9.714117107	-2.234117107	-2.386783715		
28	4	10.26780121	-1.187801208	-1.268968655		
29	5	9.96019893	-0.13019893	-0.13909597		
30	6	9.929438702	0.160561298	0.171533126		
31	7	10.51388303	0.496116969	0.530018725		
32	8	10.88300577	0.606994235	0.648472701		
33	9	11.0368069	1.033193095	1.10379552		
34	10	11.8058126	0.744187399	0.795040851		
35	11	11.22136827	0.698631728	0.746372169		
36	12	9.898678474	0.371321526	0.396695485		
37	13	11.77505237	0.024947627	0.026652403		
38	14	13.19002285	-1.040022854	-1.111091984		
39	15	9.837158019	-0.197158019	-0.210630654		

FIGURE 11.12   Excel output for single package delivery store.

Table 11.6 Customers and sales for a period of 15 consecutive weeks.

Week	Customers	Sales ($000)
1	794	9.33
2	799	8.26
3	837	7.48
4	855	9.08
5	845	9.83
6	844	10.09
7	863	11.01
8	875	11.49
9	880	12.07
10	905	12.55
11	886	11.92
12	843	10.27
13	904	11.80
14	950	12.15
15	841	9.64

We note from Figure 11.12 that r^2 is .657, indicating that 65.7% of the variation in sales can be explained by variation in the number of customers. In addition, the Y intercept, b_0, is -16.032, while the slope, b_1, is .03076. However, before we can accept the validity of this model, we must undertake proper analyses of the residuals. Since the data have been collected over a consecutive period of 15 weeks, the residuals should be plotted over time to see whether a pattern exists. Figure 11.13 represents such a plot for the 15-week sales data.

From Figure 11.13, we observe that the points tend to fluctuate up and down in a cyclical pattern. This cyclical pattern would give us strong cause for concern about the autocor-

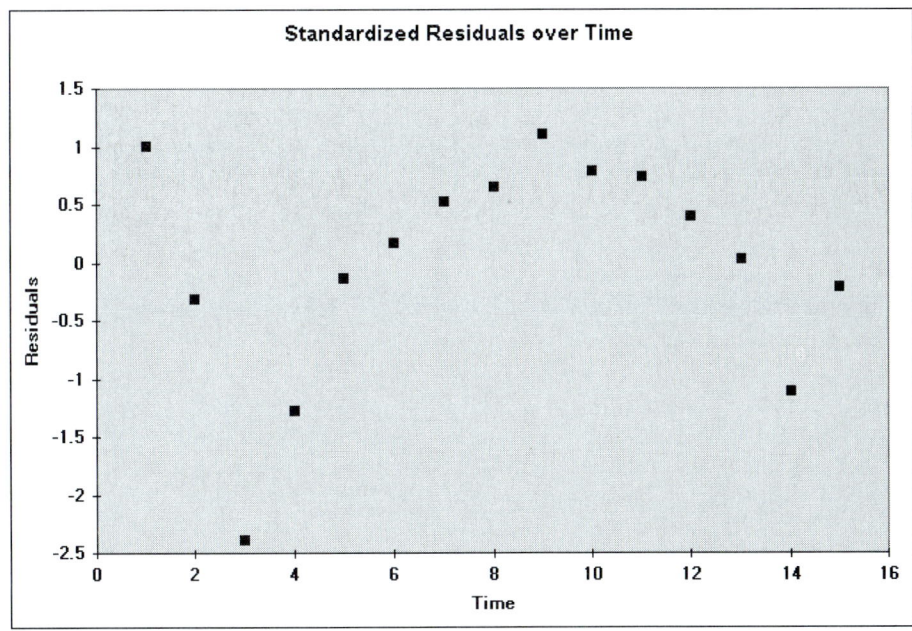

FIGURE 11.13 Excel plot of residuals over time for the 15-week sales data.

relation of the residuals and, hence, a violation in the assumption of independence of the residuals.

11.11.3 The Durbin-Watson Procedure

In addition to residual plots, autocorrelation can also be detected and measured by using the **Durbin-Watson statistic**. This statistic measures the correlation of each residual and the residual for the time period immediately preceding the one of interest. The Durbin-Watson statistic (D) is defined as follows:

$$D = \frac{\sum\limits_{i=2}^{n}(e_i - e_{i-1})^2}{\sum\limits_{i=1}^{n}e_i^2} \tag{11.16}$$

where e_i = residual at the time period i

Although the computation of the Durbin-Watson statistic can be obtained using a Microsoft Excel formula, (see Figure 11.10.Excel on page 573), for illustrative purposes, the computations for the 15-week sales data are summarized in Table 11.7.

To better understand what the Durbin-Watson statistic is measuring, we need to examine the composition of the D statistic presented in Equation (11.16). The numerator $\sum\limits_{i=2}^{n}(e_i - e_{i-1})^2$ represents the squared difference in two successive residuals, summed from the second observation to the nth observation. The denominator represents the sum of the

Table 11.7 Computation of the Durbin-Watson statistic for the regression analysis of the single package delivery store.

Week	Sales (Y_i)	$\hat{Y}_i$	$e_i = Y_i - \hat{Y}_i$	e_{i-1}	$(e_i - e_{i-1})$	$(e_i - e_{i-1})^2$	e_i^2
1	9.33	8.3914	0.93857	*	*	*	0.88092
2	8.26	8.5452	−0.28523	0.93857	−1.22380	1.49769	0.08136
3	7.48	9.7141	−2.23412	−0.28523	−1.94889	3.79817	4.99128
4	9.08	10.2678	−1.18780	−2.23412	1.04632	1.09478	1.41087
5	9.83	9.9602	−0.13020	−1.18780	1.05760	1.11852	0.01695
6	10.09	9.9294	0.16056	−0.13020	0.29076	0.08454	0.02578
7	11.01	10.5139	0.49612	0.16056	0.33556	0.11260	0.24613
8	11.49	10.8830	0.60699	0.49612	0.11088	0.01229	0.36844
9	12.07	11.0368	1.03319	0.60699	0.42620	0.18165	1.06749
10	12.55	11.8058	0.74419	1.03319	−0.28901	0.08352	0.55381
11	11.92	11.2214	0.69863	0.74419	−0.04556	0.00208	0.48809
12	10.27	9.8987	0.37132	0.69863	−0.32731	0.10713	0.13788
13	11.80	11.7751	0.02495	0.37132	−0.34637	0.11997	0.00062
14	12.15	13.1900	−1.04002	0.02495	−1.06497	1.13416	1.08165
15	9.64	9.8372	−0.19716	−1.04002	0.84286	0.71042	0.03887

$$\sum_{i=2}^{n}(e_i - e_{i-1})^2 = 10.058 \qquad \sum_{i=1}^{n}e_i^2 = 11.39$$

squared residuals, $\sum_{i=1}^{n} e_i^2$. When successive residuals are positively autocorrelated, the value of D will approach 0. If the residuals are not correlated, the value of D will be close to 2. (If there is negative autocorrelation, which rarely happens, D will be greater than 2 and could even approach its maximum value of 4.)

For the data in Table 11.7, we use Equation (11.16) and obtain

$$D = \frac{10.058}{11.39} = .883$$

The crux of the issue in using the Durbin-Watson statistic is the determination of when the autocorrelation is large enough to make the D statistic fall sufficiently below 2 to cause concern about the validity of the model. The answer to this question is dependent on the number of observations being analyzed and the number of independent variables in the model (in simple linear regression, $p = 1$). Table 11.8 has been extracted from Appendix E, Table E.9, the table of the Durbin-Watson statistic.

From Table 11.8, we observe that two values are shown in the table for each combination of level of significance (α), n (sample size), and p (the number of independent variables in the model). The first value, d_L, represents the lower critical value when there is no autocorrelation in the data. If D is below d_L, we may conclude that there is evidence of autocorrelation among the residuals. Under such a circumstance, the least squares methods that we have considered in this chapter are inappropriate and alternative methods need to be used (see References 5 and 12). The second value, d_U, represents the upper critical value of D above which we would conclude that there is no evidence of autocorrelation among the residuals. If D is between d_L and d_U, we are unable to make a definite conclusion.

Thus, as illustrated in Table 11.8, for our data concerning the single package delivery store, with one independent variable ($p = 1$), and 15 observations ($n = 15$), $d_L = 1.08$ and $d_U = 1.36$. Since $D = 0.883 < 1.08$, we may conclude that there is autocorrelation among the residuals. Our analysis of the data of Figure 11.12, using the least squares method was inappropriate due to the presence of serious autocorrelation among the residuals. We need to consider the alternative approaches discussed in Reference 12.

Table 11.8   **Finding critical values of the Durbin-Watson statistic.**

| | $\alpha = .05$ | | | | | | | | | |
| | $p = 1$ | | $p = 2$ | | $p = 3$ | | $p = 4$ | | $p = 5$ | |
n	d_L	d_U	d_L	d_U	d_L	d_U	d_L	d_U	d_L	d_U
15	1.08	1.36	.95	1.54	.82	1.75	.69	1.97	.56	2.21
16	1.10	1.37	.98	1.54	.86	1.73	.74	1.93	.62	2.15
17	1.13	1.38	1.02	1.54	.90	1.71	.78	1.90	.67	2.10
18	1.16	1.39	1.05	1.53	.93	1.69	.82.	1.87	.71	2.06
19	1.18	1.40	1.08	1.53	.97	1.68	.86	1.85	.75	2.02
⋮	⋮	⋮	⋮	⋮	⋮	⋮	⋮	⋮	⋮	⋮
90	1.63	1.68	1.61	1.70	1.59	1.73	1.57	1.75	1.54	1.78
95	1.64	1.69	1.62	1.71	1.60	1.73	1.58	1.75	1.56	1.78
100	1.65	1.69	1.63	1.72	1.61	1.74	1.59	1.76	1.57	1.78

Note: n = number of observations; p = number of independent variables.
Source: Table E.9.

11.11.4 Using Microsoft Excel to Study Autocorrelation

11-11-4.XLS

We have used two approaches in this section to evaluate the autocorrelation of data collected over time—residual plots and the Durbin-Watson statistic. To use Microsoft Excel for these approaches, we would first use the Data Analysis regression tool, discussed in Section 11.7, to produce the residuals required for these analyses.

To generate a residual plot, the Chart Wizard could be used to produce a *XY* Plot of the residuals over time (the observation column), following a procedure similar to the one described in Section 11.6.3. This will produce a plot like the one illustrated in Figure 11.3.

To generate the Durbin-Watson statistic requires designing a sheet such as the Calculations sheet in Table 11.3.Excel that contains formulas that use the SUMXMY2 and the SUMSQ Excel functions.

The SUMXMY2 function computes the sum of the squared differences between a set of *X* and *Y* values $\sum_{i=1}^{n}(X_i - Y_i)^2$. The format of the function is

$$=SUMXMY2(\textit{range of X variable, range of Y variable})$$

The SUMSQ function computes the sum of the squared values of a variable, $\sum_{i=1}^{n}X_i^2$. The format of the function is

$$=SUMSQ(\textit{range of the X variable})$$

Examining Equation (11.16) on page 570 for the Durbin-Watson statistic, we can note that the numerator is the sum of the squared differences between the successive residuals e_i and e_{i-1}, while the denominator is the sum of the squared residuals e_i^2.

To implement this sheet, open the 11-11-4.XLS workbook, that contains the data of Table 11.6 on page 569 and a Regression sheet that contains the output produced by using the Data Analysis regression tool. Insert a new sheet and rename it Calculations. For this example, copy the residuals from the range C25:C39 on the Regression sheet to the range A4:A18 on the Calculations sheet.

Then enter the formula =SUMXMY2(A5:A18,A4:A17) in cell D3 to compute the sum of the squared differences between the successive residuals e_i and e_{i-1}. Next enter the formula =SUMSQ(A4:A18) in cell D4 to compute the sum of the squared residuals e_i^2. Finally, enter the formula =D3/D4 in cell D5 to compute the Durbin-Watson statistic. The results obtained for the data of Table 11.6 are displayed in Figure 11.10.Excel on page 573.

Table 11.3.Excel   **Design of the Calculations sheet for computing the Durbin-Watson statistic.**

	A	B	C	D
1	Calculations for the Durbin-Watson Statistic			
2				
3	Residuals		Squared Difference of Residuals	=SUMXMY2(A5:A18,A4:A17)
4	x.xxx		Squared Residuals	=SUMSQ(A4:A18)
5	x.xxx		Durbin-Watson Statistic	=D3/D4
	.			
	.			
	.			
18	x.xxx			

	A	B	C	D
1	Calculations for the Durbin-Watson Statistic			
2				
3	Residuals		Squared Difference of Residuals	10.05752
4	0.938573		Squared Residuals	11.39014
5	-0.28523		Durbin-Watson Statistic	0.883003
6	-2.23412			
7	-1.1878			
8	-0.1302			
9	0.160561			
10	0.496117			
11	0.606994			
12	1.033193			
13	0.744187			
14	0.698632			
15	0.371322			
16	0.024948			
17	-1.04002			
18	-0.19716			

FIGURE 11.10.EXCEL Using Microsoft Excel to calculate the Durbin-Watson statistic for the sales data.

Problems for Section 11.10

11.36 Under what circumstances would it be important to compute the Durbin-Watson statistic? Explain.

• 11.37 Referring to Problem 11.1 (pet food sales) on page 536, is it necessary to compute the Durbin-Watson statistic? Explain.

11.38 Referring to Problem 11.1, under what circumstances would it be necessary to obtain the Durbin-Watson statistic before proceeding with the least squares method of regression analysis?

PETFOOD.TXT

RESID1.TXT

11.39 Suppose the residuals for a set of data collected over 10 consecutive time periods are as follows:

Time Period	Residual
1	−5
2	−4
3	−3
4	−2
5	−1
6	+1
7	+2
8	+3
9	+4
10	+5

(a) Plot the residuals over time. What conclusions can you reach about the pattern of the residuals over time?

(b) Compute the Durbin-Watson statistic.

(c) Based on (a) and (b), what conclusion can you reach about the autocorrelation of the residuals?

11.40 Suppose that the residuals for a set of data collected over 15 consecutive time periods are as follows:

RESID2.TXT

Time Period	Residual
1	+4
2	−6
3	−1
4	−5
5	+2
6	+5
7	−2
8	+7
9	+6
10	−3
11	+1
12	+3
13	0
14	−4
15	−7

(a) Plot the residuals over time. What conclusions can you reach about the pattern of the residuals over time?

(b) Compute the Durbin-Watson statistic. At the .05 level of significance, is there evidence of positive autocorrelation among the residuals?

(c) Based on (a) and (b), what conclusion can you reach about the autocorrelation of the residuals?

11.12 CONFIDENCE INTERVAL ESTIMATE FOR PREDICTING μ_{YX}

11.12.1 Obtaining the Confidence Interval Estimate

In Sections 11.1–11.7, we were concerned with the use of regression and correlation solely for the purpose of description. The least squares method has been utilized to determine the regression coefficients and to predict the value of Y from a given value of X. In addition, the standard error of the estimate has been discussed along with the coefficients of correlation and determination.

Now that we have used residual analysis in Section 11.10 to assure ourselves that the assumptions of the least squares regression model have not been violated and that the straight-line model is appropriate, we may concern ourselves with making inferences about the relationship between the variables in a population based on our sample results. In this section we will discuss methods of making predictive inferences about the mean of Y, and in the following section we will predict an individual response value Y_I.

You may recall that in Section 11.4 the fitted regression equation was used to make predictions about the value of Y for a given X. In the package delivery sales example, we predicted that the average weekly sales for stores with 600 customers would be 7.661 (thousands of dollars). This estimate, however, is merely a point estimate of the population average value. In Chapter 6, we developed the concept of the confidence interval as an estimate of the population average. In a similar fashion, a **confidence interval estimate for the mean response** can now be developed to make inferences about the average predicted value of Y:

$$\hat{Y}_i \pm t_{n-2}\, S_{YX} \sqrt{h_i} \tag{11.17}$$

where

$$h_i = \frac{1}{n} + \frac{(X_i - \overline{X})^2}{\displaystyle\sum_{i=1}^{n}(X_i - \overline{X})^2}$$

$\hat{Y}_i$ = predicted value of Y; $\hat{Y}_i = b_0 + b_1 X_i$

S_{YX} = standard error of the estimate

n = sample size

X_i = given value of X

An examination of Equation (11.17) indicates that the width of the confidence interval is dependent on several factors. For a given level of confidence, increased variation around the line of regression, as measured by the standard error of the estimate, results in a wider interval. However, as would be expected, increased sample size reduces the width of the interval. In addition, the width of the interval also varies at different values of X. When predicting Y for values of X close to $\overline{X}$, the interval is much narrower than for predictions for X values more distant from the mean. This effect can be seen from the square root portion of Equation (11.17) and from Figure 11.14.

As displayed in Figure 11.14, the interval estimate of the true mean of Y varies *hyperbolically* as a function of the closeness of the given X to $\overline{X}$. When predictions are to be made for X values that are distant from the average value of X, the much wider interval is the trade-off for predicting at such values of X. Thus, as depicted in Figure 11.14, we observe a *confidence band effect* for the predictions.

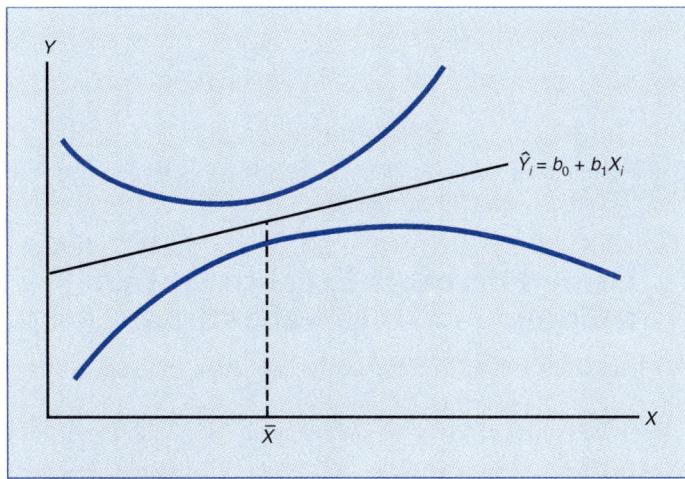

FIGURE 11.14   Interval estimates of μ_{YX} for different values of X.

Let us now use Equation (11.17) for our package delivery sales example. Suppose we desire a 95% confidence interval estimate of the true average weekly sales for all stores with 600 customers. We compute the following:

$$\hat{Y}_i = 2.423 + .00873\, X_i$$

and for $X_i = 600$, we obtain $\hat{Y}_i = 7.661$.

Also,

$$\overline{X} = 731.15 \qquad S_{YX} = .497$$

$$\sum_{i=1}^{n} (X_i - \overline{X})^2 = \sum_{i=1}^{n} X_i^2 - n\overline{X}^2 = 11{,}306{,}209 - 20(731.15)^2 = 614{,}603$$

From Table E.3, $t_{18} = 2.1009$. Thus,

$$\hat{Y}_i \pm t_{n-2}\, S_{YX} \sqrt{h_i}$$

where
$$h_i = \frac{1}{n} + \frac{(X_i - \overline{X})^2}{\displaystyle\sum_{i=1}^{n}(X_i - \overline{X})^2}$$

so that we have

$$\hat{Y}_i \pm t_{n-2}\, S_{YX} \sqrt{\frac{1}{n} + \frac{(X_i - \overline{X})^2}{\displaystyle\sum_{i=1}^{n}(X_i - \overline{X})^2}}$$

and

$$7.661 \pm (2.1009)(.497)\sqrt{\frac{1}{20} + \frac{(600 - 731.15)^2}{614{,}603}}$$

$$= 7.661 \pm (1.044)\sqrt{\frac{1}{20} + \frac{(-131.15)^2}{614{,}603}}$$

$$= 7.661 \pm (1.044)\sqrt{.078}$$

$$= 7.661 \pm .292$$

so

$$7.369 \le \mu_{YX} \le 7.953$$

Therefore, our estimate is that the average weekly sales is between 7.369 and 7.953 (thousand dollars) for stores with 600 customers.

11.12.2 Using Microsoft Excel to Obtain a Confidence Interval Estimate for μ_{YX}

11-12-2.XLS

In this section we have learned how to compute the confidence interval estimate for the average predicted value μ_{YX}. Although the Data Analysis tool does not produce this estimate, we can use results from the output of that tool and combine them with Excel formulas to compute it. Examining Equation (11.17) on page 575, recall that four quantities are required for this calculation, the predicted Y value $\hat{Y}_i$, the critical value of t, the standard error of the estimate S_{YX}, and the h_i statistic. Table 11.4.Excel presents the design for a Calculations sheet that

Table 11.4.Excel **Design for the Calculations sheet to obtain confidence and prediction limits.**

	A	B
1	Calculating the Confidence Interval Estimate	
2		
3	No. of Customers	xxx
4	No. of Observations	xx
5	Degrees of Freedom	=B4–2
6	t value	=TINV(0.05,B5)
7	Mean Number of Customers	=AVERAGE(Data!$B$2:$B$21)
8	Sum of Squared Differences	=SUM(Data!D2:D21)
9	Standard Error of the Estimate	xxx
10	h_i statistic	=1/B4+(B3–B7)^2/B8
11		
12	Predicted Sales	=TREND(Data!C2:C21,Data!B2:B21,B3)
13	Half-width	=B6*B9*SQRT(B10)
14	Lower Confidence Limit	=B12–B13
15	Upper Confidence Limit	=B12+B13
16	Half-width	=B6*B9*SQRT(1+B10)
17	Lower Prediction Limit	=B12–B16
18	Upper Prediction Limit	=B12+B16

computes confidence limits and prediction limits (discussed in Section 11.13) using the four quantities.

To implement rows 1 through 15 of this sheet, open the PACKAGE.XLS workbook, insert a new sheet and rename it Calculations. Enter the number of customers in cell B3, the number of observations in cell B4, and the standard error of the estimate in cell B9. The latter two values can be copied from the Regression Statistics table of the SUMMARY OUTPUT portion of the output produced by the Data Analysis regression tool. For the package delivery data, these values are 600, 20, and .5014952 respectively. Then do the following:

❶ Enter the formula =B4–2 to calculate the degrees of freedom in cell B5.

❷ Enter the formula =TINV(0.05,B5) to calculate the critical t statistic, as was done previously in Chapter 6, in cell B6.

❸ Calculate the average number of customers ($\overline{X}$) by entering the formula =AVERAGE(Data!$B$2:$B$21) in cell B7.

❹ Click on the Data sheet tab to make that sheet active. Calculate $(X_i - \overline{X})^2$ for each observation by entering the formula =(B2–Calculations!$B$7)^2 in cell D2 of that sheet and then copying the formula down to row 21.

❺ Click on the Calculations sheet tab to make this sheet active again. Enter the formula =SUM(Data!D2:D21) in cell B8 to sum the squared differences $\sum_{i=1}^{n}(X_i - \overline{X})^2$ entered in the Data sheet in the previous step.

❻ Enter the formula =1/B4+(B3–B7)^2 in cell B10 to calculate the h_i statistic.

❼ Enter the formula =TREND(Data!C2:C21,Data!B2:B21,B3) in cell B12 to compute the predicted sales as discussed in Section 11.6.3.

❽ Enter the formula =B6*B9*SQRT(B10) to compute the half-width of the confidence interval, equal to the $t S_{YX} \sqrt{h_i}$ term.

⑨ Enter the formula for the confidence limits, =B12–B13 and =B12+B13 in cells B14 and B15, respectively.

Figure 11.11.Excel shows the implemented Calculations sheet. (Section 11.13.2 discusses the implementation of rows 16-18 of this Calculations sheet.) This sheet computes the lower confidence limit to be 7.3664182 and the upper confidence limit to be 7.9548764. Differences between these results and the results obtained in Section 11.12.1 are due to rounding errors introduced into those calculations.

	A	B
1	Calculating the Confidence Interval Estimate	
2		
3	No. of Customers	600
4	No. of Observations	20
5	Degrees of Freedom	18
6	t value	2.1009237
7	Mean No. of Customers	731.15
8	Sum of Squared Difference	614602.55
9	Standard Error of the Estimate	0.5014952
10	Hi Statistic	0.0779861
11		
12	Predicted Sales	7.6606473
13	Half-width	0.2942291
14	Lower Confidence Limit	7.3664182
15	Upper Confidence Limit	7.9548764
16	Half-width	1.0939152
17	Lower Prediction Limit	6.5667321
18	Upper Prediction Limit	8.7545625

FIGURE 11.11.EXCEL
Confidence and prediction limits obtained from Excel for the package delivery sales data.

Problems for Section 11.12

Note: *The problems in this section can be solved using Microsoft Excel.*

PETFOOD.TXT

S-SITE.XLS

WORKHRS.TXT

TOMYIELD.TXT

LIMO.TXT

● 11.41 Referring to the pet food sales problem (pages 536, 544, and 546), set up a 90% confidence interval estimate of the average weekly sales for all stores that have 8 feet of shelf space for pet food.

11.42 Referring to the site selection problem (pages 536, 544, and 547), set up a 95% confidence interval estimate of the average sales for stores with 4,000 square feet.

● 11.43 Referring to the production worker-hours problem (pages 537, 544, and 547), set up a 90% confidence interval estimate of the average worker hours for all production runs with a lot size of 45.

11.44 Referring to the tomato yield problem (pages 537, 544, and 547), set up a 90% confidence interval estimate of the average yield for all tomatoes that have been fertilized with 15 pounds per 100 square feet of natural organic fertilizer.

11.45 Referring to the airport travel problem (pages 537, 544, and 547), set up a 95% confidence interval estimate of the average travel time for all distances of 21 miles.

11.13 PREDICTION INTERVAL ESTIMATE FOR AN INDIVIDUAL RESPONSE Y_I

11.13.1 Obtaining the Prediction Interval Estimate

In addition to the need to obtain a confidence interval estimate for the average value, it is often important to be able to predict the response that would be obtained for an individual value. Although the form of the prediction interval estimate is similar to the confidence interval estimate of Equation (11.17), the prediction interval is estimating an individual value, not a parameter. Thus, the **prediction interval for an individual response** Y_I at a particular value X_i is provided in Equation (11.18).

$$\hat{Y}_i \pm t_{n-2}\, S_{YX}\sqrt{1 + h_i} \qquad (11.18)$$

where h_i, $\hat{Y}_i$, S_{YX}, n, and X_i are defined as in Equation (11.17) on page 000.

Suppose we desire a 95% prediction interval estimate of the weekly sales for an individual store with 600 customers. We compute the following:

$$\hat{Y}_i = 2.423 + .00873X_i$$

and for $X_i = 600$, $\hat{Y}_i = 7.661$.

Also, from Section 11.12.1,

$$\bar{X} = 731.15, \qquad S_{YX} = .497, \qquad \text{and } h_i = .078$$

From Table E.3, $t_{18} = 2.1009$. Thus, from Equation (11.18)

$$\hat{Y}_i \pm t_{n-2}\, S_{YX}\sqrt{1 + h_i}$$

so that

$$7.661 \pm (2.1009)(.497)\sqrt{1 + .078}$$
$$= 7.661 \pm (1.044)\sqrt{1.078}$$
$$= 7.661 \pm 1.084$$

so

$$6.577 \le Y_I \le 8.745$$

Therefore, with 95% confidence, our estimate is that the weekly sales for an individual store that has 600 customers is between 6.577 and 8.745 (thousand dollars). We note that this prediction interval is much wider than the confidence interval estimate obtained in Section 11.12 for the average value.

11.13.2 Using Microsoft Excel to Obtain a Prediction Interval Estimate for Y_I

Now that we have developed the prediction interval estimate for the average predicted value Y_I, we can continue to implement the Calculations sheet presented in Table 11.4.Excel on page 000. Examining Equation (11.18) on page 577, we may observe that the only difference between this equation and Equation (11.17) for the confidence limits for μ_{YX} is that there is an additional value of 1 under the square root. Thus, to obtain the prediction limits, we can enter the formula =B6*B9*SQRT(1+B10) in cell B16 and the formulas =B12–B16 and =B12+B16 in cells B17 and B18, respectively.

The results are illustrated in Figure 11.11.Excel on page 578. Once again, any minor differences from the results obtained in Section 11.13.1 are due to rounding errors.

Problems for Section 11.13

Note: *The problems in this section can be solved using Microsoft Excel.*

PETFOOD.TXT

S-SITE.XLS

WRKHRS.TXT

TOMYIELD.TXT

LIMO.TXT

● 11.46 Referring to Problem 11.41 (pet food sales) on page 578
    (a) Set up a 90% prediction interval of the weekly sales of an individual store that has 8 feet of shelf space for pet food.
    (b) Explain the difference in the results obtained in (a) and those in Problem 11.41.

11.47 Referring to Problem 11.42 (site selection) on page 578
    (a) Set up a 95% prediction interval of the sales of an individual store that has 4,000 square feet.
    (b) Explain the difference in the results obtained in (a) and those in Problem 11.42.

● 11.48 Referring to Problem 11.43 (production worker-hours) on page 578, set up a 90% prediction interval of the number of worker-hours for a single lot size of 45.

11.49 Referring to Problem 11.44 (tomato yield) on page 578, set up a 90% prediction interval of the yield of tomatoes for an individual plot that has been fertilized with 25 pounds per 100 square feet of natural organic fertilizer.

11.50 Referring to Problem 11.45 (airport travel) on page 578, set up a 95% prediction interval of the travel time for an individual trip of 21 miles.

11.14   INFERENCES ABOUT THE POPULATION PARAMETERS IN REGRESSION AND CORRELATION

In the preceding two sections we used statistical inference to develop a confidence interval estimate for μ_{YX}, the true mean value of Y, and a prediction interval for Y_I, an individual observation. In this section, statistical inference will be used to draw conclusions about the population slope β_1 and the population correlation coefficient ρ.

We can determine whether a significant relationship between the variables X and Y exists by testing whether β_1 (the true slope) is equal to 0. If this hypothesis is rejected, one could conclude that there is evidence of a linear relationship. The null and alternative hypotheses could be stated as follows:

$$H_0: \beta_1 = 0 \qquad \text{(There is no relationship.)}$$

$$H_1: \beta_1 \neq 0 \qquad \text{(There is a relationship.)}$$

and the test statistic for this is given by

The *t* statistic equals the difference between the sample slope and population slope divided by the standard error of the slope.

$$t = \frac{b_1 - \beta_1}{S_{b_1}} \qquad (11.19)$$

where
$$S_{b_1} = \frac{S_{YX}}{\sqrt{\sum_{i=1}^{n} X_i^2 - n\overline{X}^2}}$$

and the test statistic t follows a t distribution with $n - 2$ degrees of freedom.

Returning to our package delivery sales example, let us now test whether the sample results enable us to conclude that a significant relationship between the number of customers and the weekly sales exists at the .05 level of significance. The results from Sections 11.4 and 11.5 gave the following information:

$$b_1 = +.00873 \qquad n = 20 \qquad S_{YX} = .497$$

$$\overline{X} = 731.15 \qquad \sum_{i=1}^{n} X_i^2 = 11,306,209$$

Therefore, to test the existence of a relationship at the .05 level of significance, we have

$$S_{b_1} = \frac{S_{YX}}{\sqrt{\sum_{i=1}^{n} X_i^2 - n\overline{X}^2}}$$

$$= \frac{.497}{\sqrt{11,306,209 - 20(731.15)^2}} = \frac{.497}{\sqrt{614,603}} = .000634$$

and, under the null hypothesis, $\beta_1 = 0$ so that

$$t = \frac{b_1}{S_{b_1}}$$

$$= \frac{.00873}{.000634} = 13.77$$

Since $t = 13.77 > t_{18} = 2.1009$, we reject H_0. Hence, we can conclude that there is a significant linear relationship between average weekly sales and the number of customers (see Figure 11.15).

A second, equivalent method for testing the existence of a linear relationship between the variables is to set up a confidence interval estimate of β_1 and to determine whether the hypothesized value ($\beta_1 = 0$) is included in the interval. The confidence interval estimate of β_1 would be obtained by using the following formula:

$$b_1 \pm t_{n-2} S_{b_1} \qquad\qquad (11.20)$$

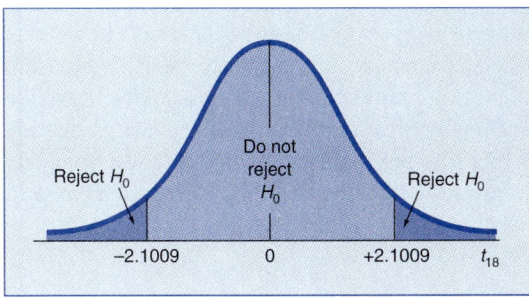

FIGURE 11.15   Testing a hypothesis about the population slope at the .05 level of significance with 18 degrees of freedom.

If there were a 95% confidence interval estimate desired here, we would have $b_1 = +.00873$, $t_{18} = 2.1009$, and $S_{b_1} = .000634$. Thus,

$$b_1 \pm t_{n-2} S_{b_1} = +.00873 \pm (2.1009)(.000634)$$

$$= +.00873 \pm .00133$$

$$+.0074 \leq \beta_1 \leq +.01006$$

From Equation (11.20), the true slope is estimated with 95% confidence to be between +.0074 and +.01006 (i.e., $7.40 to $10.06). Since these values are above 0, we can conclude that there is a significant linear relationship between weekly sales and number of customers. On the other hand, had the interval included 0, no relationship would have been determined.

A third method for examining the existence of a linear relationship between two variables involves the sample correlation coefficient r. The existence of a relationship between X and Y, which was tested using Equation (11.19), could be tested in terms of the correlation coefficient with equivalent results. Testing for the existence of a linear relationship between two variables is the same as determining whether there is any significant correlation between them. The population correlation coefficient ρ is hypothesized as equal to 0. Thus, the null and alternative hypotheses would be

$$H_0: \rho = 0 \qquad \text{(There is no correlation.)}$$

$$H_1: \rho \neq 0 \qquad \text{(There is correlation.)}$$

The test statistic for determining the existence of a significant correlation is given by

$$t = \frac{r - \rho}{\sqrt{\dfrac{1 - r^2}{n - 2}}} \qquad (11.21)$$

where the test statistic t follows a t distribution with $n - 2$ degrees of freedom.

To demonstrate that this statistic produces the same result as the test for the existence of a slope [Equation (11.19)], we will use the package delivery sales data. For these data, $r = +.956$, $r^2 = .913$, and $n = 20$, so testing the null hypothesis, we have

$$t = \frac{r}{\sqrt{\dfrac{1 - r^2}{n - 2}}}$$

$$= \frac{.956}{\sqrt{\dfrac{1 - .913}{20 - 2}}} = 13.75$$

We may note that this t value is, except for possible rounding error, the same as that obtained by using Equation (11.19). Therefore, in a linear regression analysis Equations (11.19) and (11.21) give equivalent alternative ways of determining the existence of a relationship between two variables. However, if the sole purpose of a particular study is to determine the existence of correlation, then Equation (11.21) is more appropriate. For instance, in Section 11.8, we studied the association of the price of a six-pack of a brand-name cola soft drink and the price of chicken. Had we wanted to determine the significance of the correlation between these two variables, we could have used Equation (11.21) as follows:

$$H_0: \rho = 0 \qquad \text{(There is no correlation.)}$$

$$H_1: \rho \neq 0 \qquad \text{(There is correlation.)}$$

If a level of significance of .05 was selected, we would have (see Figure 11.16)

$$t = \frac{r}{\sqrt{\dfrac{1 - r^2}{n - 2}}}$$

$$= \frac{.883}{\sqrt{\dfrac{1 - (.883)^2}{9 - 2}}} = \frac{.883}{.1774} = +4.98$$

Since $t = 4.98 > t_7 = 2.3646$, we reject H_0. Since the null hypothesis has been rejected, we would conclude that there is evidence of an association between the price of the brand-name cola soft drink and the price of chicken.

When inferences concerning the population slope were discussed, confidence intervals and tests of hypothesis were used interchangeably. However, when examining the correlation coefficient, the development of a confidence interval becomes more complicated because the shape of the sampling distribution of the statistic r varies for different values of the true correlation coefficient. Methods for developing a confidence interval estimate for the correlation coefficient are presented in Reference 12.

Now that we have discussed the test of hypothesis and the confidence interval estimate for the slope, we can return to the Excel output of Figure 11.8.Excel on page 556. In the third section of the text output in Panel A, next to the Y intercept and the slope, is a column labeled Standard Error that provides the standard error of each of the regression coefficients. As indicated, the standard error of the slope is 0.00063969. In the next column, the t statistic for testing the slope equal to 13.6461991 is provided followed in the next column by the p-value of .000000000062062 (expressed in scientific notation as 6.2062E-11). In the last two columns, Microsoft Excel provides the lower and upper confidence interval estimates for the slope. Once again, any minor differences from the results obtained in Section 11.14 are due to rounding errors.

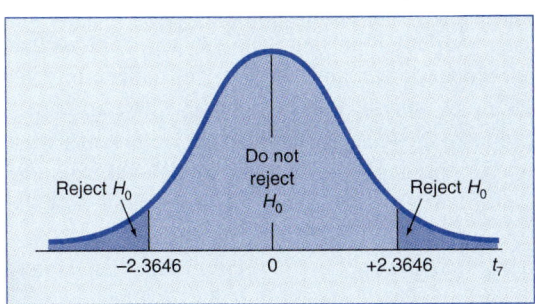

FIGURE 11.16 Testing for the existence of correlation at the .05 level of significance with 7 degrees of freedom.

Problems for Section 11.14

Note: *The problems in this section can be solved using Microsoft Excel.*

- 11.51 Referring to the pet food sales problem (pages 536, 544, and 546), at the .05 level of significance, is there evidence of a linear relationship between shelf space and sales?

11.52 Referring to the site selection problem (pages 536, 544, and 547), at the .05 level of significance, is there evidence of a linear relationship between annual sales and square footage?

PETFOOD.TXT

S-SITE.XLS

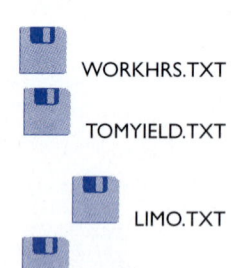

WORKHRS.TXT

TOMYIELD.TXT

LIMO.TXT

SUPERMKT.TXT

CITY.TXT

● 11.53 Referring to the production worker-hours problem (pages 537, 544, and 547), at the .01 level of significance, is there evidence of a linear relationship between lot size and worker-hours?

11.54 Referring to the tomato yield problem (pages 537, 544, and 547), at the .01 level of significance, is there evidence of a linear relationship between the amount of fertilizer used and the yield of tomatoes?

11.55 Referring to the airport travel problem (pages 537, 544, and 547), at the .05 level of significance, is there evidence of a relationship between distance and travel time?

11.56 Referring to Problem 11.29 on page 560, at the .01 level of significance, is there evidence of a linear relationship between the price of a six-pack of brand-name cola soft drink and 100 tablets of a pain reliever?

11.57 Referring to Problem 11.30 on page 561, at the .05 level of significance, is there evidence of a linear relationship between the cost of a women's haircut and a men's dress shirt?

11.15 PITFALLS IN REGRESSION AND ETHICAL ISSUES

11.15.1 Introduction

Regression and correlation analysis are perhaps the most widely used and, unfortunately, the most widely misused statistical techniques that are applied to business and economics. The difficulties frequently come from the following sources:

1. Lacking an awareness of the assumptions of least squares regression
2. Knowing how to evaluate the assumptions of least squares regression
3. Knowing what the alternatives to least squares regression are if a particular assumption is violated
4. Thinking that correlation implies causation
5. Using a regression model without knowledge of the subject matter

11.15.2 The Pitfalls of Regression

The widespread availability of spreadsheet and statistical software has removed the computational block that prevented many users from applying regression analysis to situations that required forecasting. With this positive development of availability comes the realization that, for many users, the access to powerful techniques has not been accompanied by an understanding of how to use regression analysis properly. How can a user be expected to know what the alternatives to least squares regression are if a particular assumption is violated, when he or she in many instances is not even aware of the assumptions of regression, let alone how the assumptions can be evaluated?

The necessity of going beyond the basic number crunching—of computing the Y intercept, the slope, and r^2—can be illustrated by referring to Table 11.9, a classical pedagogical piece of statistical literature that deals with the importance of observation through scatter plots and residual analysis.

Anscombe (Reference 2) showed that for the four data sets given in Table 11.9, the following results would be obtained:

$$\hat{Y}_i = 3.0 + .5X_i$$

$$S_{YX} = 1.236$$

$$S_{b_i} = .118$$

$$r^2 = .667$$

Table 11.9 **Four sets of artificial data.**

Data Set A		Data Set B		Data Set C		Data Set D	
X_i	Y_i	X_i	Y_i	X_i	Y_i	X_i	Y_i
10	8.04	10	9.14	10	7.46	8	6.58
14	9.96	14	8.10	14	8.84	8	5.76
5	5.68	5	4.74	5	5.73	8	7.71
8	6.95	8	8.14	8	6.77	8	8.84
9	8.81	9	8.77	9	7.11	8	8.47
12	10.84	12	9.13	12	8.15	8	7.04
4	4.26	4	3.10	4	5.39	8	5.25
7	4.82	7	7.26	7	6.42	19	12.50
11	8.33	11	9.26	1	7.81	8	5.56
13	7.58	13	8.74	13	12.74	8	7.91
6	7.24	6	6.13	6	6.08	8	6.89

Source: F. J. Anscombe, "Graphs in Statistical Analysis," *American Statistician*, Vol. 27 (1973), pp. 17–21.

$$SSR = \text{explained variation} = \sum_{i=1}^{n} (\hat{Y}_i - \bar{Y})^2 = 27.50$$

$$SSE = \text{unexplained variation} = \sum_{i=1}^{n} (Y_i - \hat{Y}_i)^2 = 13.75$$

$$SST = \text{total variation} = \sum_{i=1}^{n} (Y_i - \bar{Y})^2 = 41.25$$

Thus, with respect to the pertinent statistics associated with a simple linear regression, the four data sets are identical. Had we stopped our analysis at this point, valuable information in the data would be lost. Table 11.10 gives the standardized residuals e_i/S_{YX} for each data set.

Table 11.10 **Standardized Residuals.**

	Data Set A	Data Set B	Data Set C	Data Set D	
X_i	e_i/S_{YX}	e_i/S_{YX}	e_i/S_{YX}	X_i	e_i/S_{YX}
4	−.599	−1.536	.314	8	−.340
5	.145	−.614	.185	8	−1.003
6	1.002	.105	.064	8	.574
7	−1.359	.614	−.065	8	1.489
8	−.041	.922	−.186	8	1.189
9	1.059	1.027	−.315	8	.032
10	.032	.922	−.437	8	−1.416
11	−.138	.614	−.558	19	.000
12	1.487	.105	−.687	8	−1.165
13	−1.554	−.614	2.622	8	.736
14	−.033	−1.536	−.937	8	−.089

Source: F. J. Anscombe, "Graphs in Statistical Analysis," *American Statistician*, Vol. 27 (1973), pp. 17–21.

When the standardized residuals are plotted[2] against $\hat{Y}$, we see how different the data sets are. Panels A, B, C, and D of Figure 11.17 graphically depict, for each data set, a plot of the standardized residuals against the fitted values $\hat{Y}$. While the plot for data set A does not show any obvious anomalies, this is not the case for data sets B, C, and D. The parabolic form of the residual plot for B probably indicates that the basic simple linear regression model should also include a curvilinear term, as will be developed in Section 12.10. The plot for data set C clearly depicts what may very well be an *outlying* observation. If this is the case, we may deem it appropriate to remove the outlier and reestimate the basic model.[3] The result of this exercise would probably be a relationship much different from what was originally uncovered. Similarly, the plot for data set D would be evaluated cautiously because the fitted model is so dependent on the outcome of a single response ($X_8 = 19$ and $Y_8 = 12.50$).

In summary, residual plots are of vital importance to a complete regression analysis. The information they provide is so basic to a credible analysis that such plots should *always* be included as part of a regression analysis. Thus, a strategy that might be employed to avoid the first three pitfalls of regression listed would involve the following approach:

1. Always start with a scatter plot to observe the possible relationship between X and Y.

2. Check the assumptions of regression after the regression model has been fit, *before* moving on to using the results of the model.

3. Plot the residuals (or standardized or Studentized residuals) versus the independent variable. This will enable you to determine whether the model fit to the data is an appropriate one and will allow you to check visually for violations of the homoscedasticity assumption.

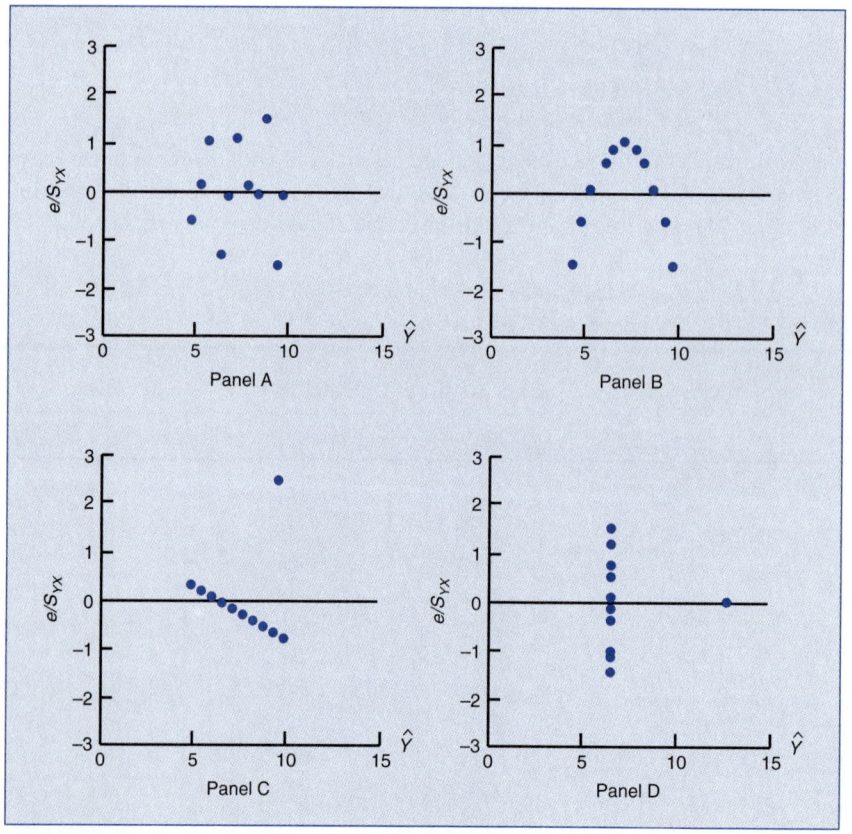

FIGURE 11.17   **Plot of $\hat{Y}_i$ versus standardized residuals.**   *Source:* F. J. Anscombe, "Graphs in Statistical Analysis," *American Statistician,* Vol. 27 (1973), pp. 17–21.

4. Use a histogram, stem-and-leaf display, box-and-whisker plot, or normal probability plot of the residuals to evaluate graphically whether the normality assumption has been seriously violated.

5. If the data have been collected in sequential order, plot the residuals in time order and compute the Durbin-Watson statistic.

6. If the evaluation done in 3–5 indicates violations in the assumptions, use alternative methods to least squares regression or alternative least squares models (curvilinear or multiple regression), depending on what the evaluation has indicated.

7. If the evaluation done in 3–5 does not indicate violations in the assumptions, then the inferential aspects of the regression analysis can be undertaken. Confidence and prediction intervals can be developed, and tests for the significance of the regression coefficients can be done.

● **Cautions**   In addition to the first three pitfalls considered above, two other pitfalls need to be mentioned. One involves the mistaken belief that correlation implies causation. In many instances, the covariation between variables is spurious in that the relationship is actually caused by a third factor that has not been or cannot be measured.

Another pitfall involves the fact that a good-fitting model does not necessarily mean that the model can be used for prediction. An individual with knowledge of the subject matter would have to be convinced that the process that produced the data will remain stable in the future in order to use the model for predictive purposes.

11.15.3   Ethical Considerations

Ethical considerations arise when a user wishing to develop forecasts manipulates the process of developing the regression model. The key here is intent. Unethical behavior occurs when someone uses regression analysis to:

1. Forecast a response variable of interest with the willful intent of possibly excluding certain variables from consideration in the model.

2. Delete observations from the model to obtain a better model without giving reasons for deleting these observations.

3. Make forecasts without providing an evaluation of the assumptions when he or she knows that the assumptions of least squares regression have been violated.

All of these situations should make us realize even more the importance of following the steps given in Section 11.15.2 and knowing the assumptions of regression, how to evaluate them, and what to do if any of them have been violated.

11.16   SUMMARY AND OVERVIEW

As seen in the chapter summary chart, we developed the simple linear regression model, discussed the assumptions of the model, and showed how these assumptions could be evaluated. To be sure you understand what has been covered, you should be able to answer the following conceptual questions:

1. What is the interpretation of the Y intercept and the slope in a regression model?

2. What is the interpretation of the coefficient of determination?

3. Why should a residual analysis always be done as part of the development of a regression model?

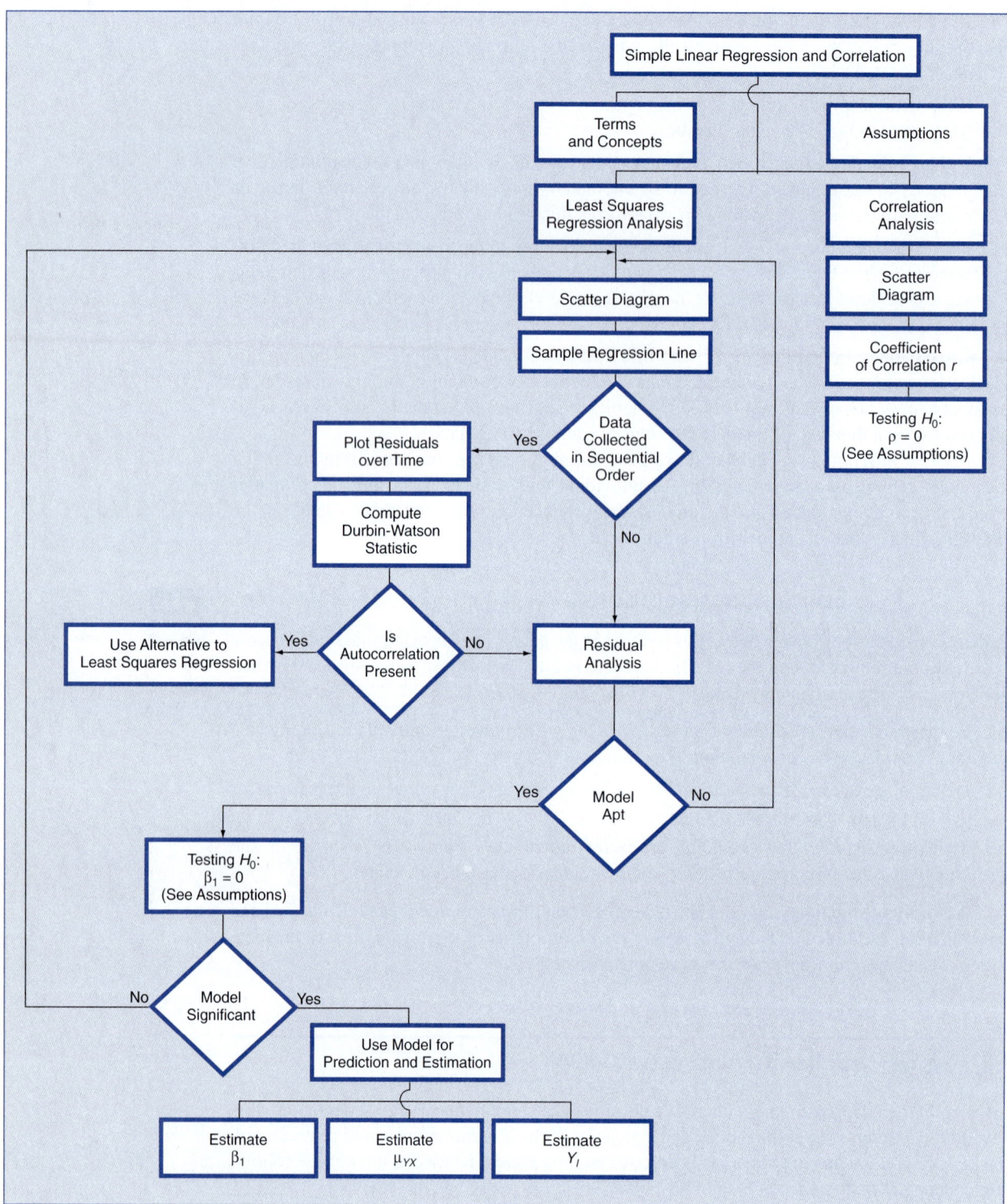

Chapter 11 summary chart.

4. What are the assumptions of regression analysis and how can they be evaluated?

5. What is the Durbin-Watson statistic and when and how should it be used in regression analysis?

6. What is the difference between a confidence interval estimate of the mean response μ_{YX} and a prediction interval estimate of Y_I?

In Chapter 12, we will continue our discussion of regression analysis by considering a variety of multiple regression models.

Getting It All Together

Key Terms

Chapter Review Problems

Note: *The Chapter Review Problems can be solved using Microsoft Excel.*

● 11.58 A statistician for an American automobile manufacturer would like to develop a statistical model for predicting delivery time (the days between the ordering of the car and the actual delivery of the car) of custom-ordered new automobiles. The statistician believes there is a linear relationship between the number of options ordered on the car and delivery time. A random sample of 16 cars is selected with the results given at the top of page 590:

DELIVERY.TXT

Relating delivery time with options ordered

Car	Options Ordered, X	Delivery Time, Y (in days)	Car	Options Ordered, X	Delivery Time, Y (in days)
1	3	25	9	12	44
2	4	32	10	12	51
3	4	26	11	14	53
4	7	38	12	16	58
5	7	34	13	17	61
6	8	41	14	20	64
7	9	39	15	23	66
8	11	46	16	25	70

(a) Set up a scatter diagram.
(b) Use the least squares method to find the regression coefficients b_0 and b_1.
(c) Interpret the meaning of the Y intercept b_0 and the slope b_1 in this problem.
(d) If a car with 16 options is ordered, how many days would you predict it would take to be delivered?
(e) Compute the standard error of the estimate.
(f) Compute the coefficient of determination r^2 and interpret its meaning in this problem.
(g) Compute the adjusted r^2 and compare it with the coefficient of determination r^2.
(h) Compute the coefficient of correlation r.
(i) Set up a 95% confidence interval estimate of the average delivery time for all cars ordered with 16 options.
(j) Set up a 95% prediction interval estimate of the delivery time for an individual car that was ordered with 16 options.
(k) At the .05 level of significance, is there evidence of a linear relationship between number of options and delivery time?
(l) Set up a 95% confidence interval estimate of the population slope.
(m) Perform a residual analysis on your results and determine the adequacy of the fit of the model.
(n) What assumptions about the relationship between the number of options and delivery time would the statistician need to make to use this regression model for predictive purposes in the future?

BET.TXT

11.59 An official of a local racetrack would like to develop a model to forecast the amount of money bet (in millions of dollars) based on attendance. A random sample of 15 days is selected with the following results:

Relating betting with attendance.

Day	Attendance (000)	Amount Bet ($000,000)	Day	Attendance (000)	Amount Bet ($000,000)
1	14.5	0.70	9	16.3	0.71
2	21.2	0.83	10	32.1	1.04
3	11.6	0.62	11	27.6	0.97
4	31.7	1.10	12	34.8	1.13
5	46.8	1.27	13	29.3	0.91
6	31.4	1.02	14	19.2	0.68
7	40.0	1.15	15	16.3	0.63
8	21.0	0.80			

Hint: Determine which are the independent and dependent variables.

(a) Set up a scatter diagram.

(b) Assuming a linear relationship, use the least squares method to find the regression coefficients b_0 and b_1.

(c) Interpret the meaning of the slope b_1 in this problem.

(d) Predict the amount bet for a day on which attendance is 20,000.

(e) Compute the standard error of the estimate.

(f) Compute the coefficient of determination r^2 and interpret its meaning in this problem.

(g) Compute the coefficient of correlation r.

(h) Compute the Durbin-Watson statistic and, at the .05 level of significance, determine whether there is any autocorrelation in the residuals.

(i) Based on the results of (h), what conclusions can you reach concerning the validity of the model fit in (b)?

(j) Set up a 95% confidence interval estimate of the average amount of money bet when attendance is 20,000.

(k) Set up a 95% prediction interval for the amount of money bet on a day in which attendance is 20,000.

(l) At the .05 level of significance, is there evidence of a linear relationship between the amount of money bet and attendance?

(m) Set up a 95% confidence interval estimate of the true slope.

(n) Discuss why you should not predict the amount bet on a day in which the attendance exceeded 46,800 or was below 11,600.

(o) Perform a residual analysis on your results and determine the adequacy of the fit of the model.

(p) Suppose that the amount bet on day 5 was 1.5 million dollars. Do (a)–(o) with these data and compare the difference in the results.

ICECREAM.TXT

11.60 The owner of a large chain of ice-cream stores would like to study the effect of atmospheric temperature on sales during the summer season. A random sample of 21 days is selected with the results given as follows:

Relating sales to temperature.

Day	Daily High Temperature (°F)	Sales per Store ($000)	Day	Daily High Temperature (°F)	Sales per Store ($000)
1	63	1.52	12	75	1.92
2	70	1.68	13	98	3.40
3	73	1.80	14	100	3.28
4	75	2.05	15	92	3.17
5	80	2.36	16	87	2.83
6	82	2.25	17	84	2.58
7	85	2.68	18	88	2.86
8	88	2.90	19	80	2.26
9	90	3.14	20	82	2.14
10	91	3.06	21	76	1.98
11	92	3.24			

Hint: Determine which are the independent and dependent variables.

(a) Set up a scatter diagram.

(b) Assuming a linear relationship, use the least squares method to find the regression coefficients b_0 and b_1.

(c) Interpret the meaning of the slope b_1 in this problem.

(d) Predict the sales per store for a day in which the temperature is 83°F.
(e) Compute the standard error of the estimate.
(f) Compute the coefficient of determination r^2 and interpret its meaning in this problem.
(g) Compute the coefficient of correlation r.
(h) Compute the adjusted r^2 and compare it with the coefficient of determination r^2.
(i) Compute the Durbin-Watson statistic and, at the .05 level of significance, determine whether there is any autocorrelation in the residuals.
(j) Based on the results of (i), what conclusions can you reach concerning the validity of the model fit in (b)?
(k) Set up a 95% confidence interval estimate of the average sales per store for all days in which the temperature is 83°F.
(l) Set up a 95% prediction interval for the sales per store on a day in which the temperature is 83°F.
(m) At the .05 level of significance, is there evidence of a linear relationship between temperature and sales?
(n) Set up a 95% confidence interval estimate of the true slope.
(o) Discuss how different your results might be if the model had been based upon temperature measured according to the Celsius (°C) scale.
(p) Perform a residual analysis on your results and determine the adequacy of the fit of the model.
(q) Suppose that the amount of sales on day 21 was 1.75. Do (a)–(p) and compare the difference in the results.

HOUSE1.TXT

11.61 Suppose we want to develop a model to predict selling price based on assessed value. A sample of 30 recently sold single-family houses in a small western city is selected to study the relationship between selling price and assessed value (the houses in the city had been reassessed at full value 1 year prior to the study). The results are as follows:

Relating selling price to assessed value.

Observation	Assessed Value ($000)	Selling Price ($000)	Observation	Assessed Value ($000)	Selling Price ($000)
1	78.17	94.10	16	84.36	106.70
2	80.24	101.90	17	72.94	81.50
3	74.03	88.65	18	76.50	94.50
4	86.31	115.50	19	66.28	69.00
5	75.22	87.50	20	79.74	96.90
6	65.54	72.00	21	72.78	86.50
7	72.43	91.50	22	77.90	97.90
8	85.61	113.90	23	74.31	83.00
9	60.80	69.34	24	79.85	97.30
10	81.88	96.90	25	84.78	100.80
11	79.11	96.00	26	81.61	97.90
12	59.93	61.90	27	74.92	90.50
13	75.27	93.00	28	79.98	97.00
14	85.88	109.50	29	77.96	92.00
15	76.64	93.75	30	79.07	95.90

Hint: First determine which are the independent and dependent variables.
(a) Plot a scatter diagram and, assuming a linear relationship, use the least squares method to find the regression coefficients b_0 and b_1.
(b) Interpret the meaning of the Y intercept b_0 and the slope b_1 in this problem.

(c) Use the regression model developed in (a) to predict the selling price for a house whose assessed value is $70,000.

(d) Compute the standard error of the estimate.

(e) Compute the coefficient of determination r^2 and interpret its meaning in this problem.

(f) Compute the coefficient of correlation r.

(g) Compute the adjusted r^2 and compare it with the coefficient of determination r^2.

(h) At the .10 level of significance, is there evidence of a linear relationship between selling price and assessed value?

(i) Set up a 90% confidence interval estimate of the average selling price for houses with an assessed value of $70,000.

(j) Set up a 90% prediction interval estimate of the selling price of an individual house with an assessed value of $70,000.

(k) Set up a 90% confidence interval estimate of the population slope.

(l) Perform a residual analysis on your results and determine the adequacy of the fit of the model.

11.62 Suppose we want to develop a model to predict assessed value based on heating area. A sample of 15 single-family houses is selected in a different city. The assessed value (in thousands of dollars) and the heating area of the houses (in thousands of square feet) are recorded with the following results:

HOUSE2.TXT

Relating assessed value to heating area.

House	Assessed Value ($000)	Heating Area of Dwelling (thousands of square feet)
1	84.4	2.00
2	77.4	1.71
3	75.7	1.45
4	85.9	1.76
5	79.1	1.93
6	70.4	1.20
7	75.8	1.55
8	85.9	1.93
9	78.5	1.59
10	79.2	1.50
11	86.7	1.90
12	79.3	1.39
13	74.5	1.54
14	83.8	1.89
15	76.8	1.59

Hint: First determine which are the independent and dependent variables.

(a) Plot a scatter diagram and, assuming a linear relationship, use the least squares method to find the regression coefficients b_0 and b_1.

(b) Interpret the meaning of the Y intercept b_0 and the slope b_1 in this problem.

(c) Use the regression model developed in (a) to predict the assessed value for a house whose heating area is 1,750 square feet.

(d) Compute the standard error of the estimate.

(e) Compute the coefficient of determination r^2 and interpret its meaning in this problem.

(f) Compute the coefficient of correlation r.

(g) Compute the adjusted r^2 and compare it with the coefficient of determination r^2.

(h) At the .10 level of significance, is there evidence of a linear relationship between assessed value and heating area?

(i) Set up a 90% confidence interval estimate of the average assessed value for houses with a heating area of 1,750 square feet.

(j) Set up a 90% prediction interval estimate of the assessed value of an individual house with a heating area of 1,750 square feet.

(k) Set up a 90% confidence interval estimate of the population slope.

(l) Perform a residual analysis on your results and determine the adequacy of the fit of the model.

(m) Suppose that the assessed value for the fourth house was 79.7. Do (a)–(l) and compare the results.

GPIGMAT.TXT

11.63 The Director of Graduate Studies at a large college of business would like to be able to predict the grade point index (GPI) of students in an MBA program based on Graduate Management Aptitude Test (GMAT) score. A sample of 20 students who had completed 2 years in the program is selected; the results are as follows:

Relating GPI to GMAT score.

Observation	GMAT Score	GPI	Observation	GMAT Score	GPI
1	688	3.72	11	567	3.07
2	647	3.44	12	542	2.86
3	652	3.21	13	551	2.91
4	608	3.29	14	573	2.79
5	680	3.91	15	536	3.00
6	617	3.28	16	639	3.55
7	557	3.02	17	619	3.47
8	599	3.13	18	694	3.60
9	616	3.45	19	718	3.88
10	594	3.33	20	759	3.76

Hint: First determine which are the independent and dependent variables.

(a) Plot a scatter diagram and assuming a linear relationship, use the least squares method to find the regression coefficients b_0 and b_1.

(b) Interpret the meaning of the Y intercept b_0 and the slope b_1 in this problem.

(c) Use the regression model developed in (a) to predict the GPI for a student with a GMAT score of 600.

(d) Compute the standard error of the estimate.

(e) Compute the coefficient of determination r^2 and interpret its meaning in this problem.

(f) Compute the coefficient of correlation r.

(g) Compute the adjusted r^2 and compare it with the coefficient of determination r^2.

(h) At the .05 level of significance, is there evidence of a linear relationship between GMAT score and GPI?

(i) Set up a 95% confidence interval estimate for the average GPI of students with a GMAT score of 600.

(j) Set up a 95% prediction interval estimate of the GPI for a particular student with a GMAT score of 600.

(k) Set up a 95% confidence interval estimate of the population slope.

(l) Perform a residual analysis on your results and determine the adequacy of the fit of the model.

(m) Suppose the GPIs of the 19th and 20th students were incorrectly entered. The GPI for student 19 should be 3.76, and the GPI for student 20 should be 3.88. Do (a)–(l) and compare the results.

11.64 The manager of the purchasing department of a large banking organization would like to develop a model to predict the amount of time it takes to process invoices. Data are collected from a sample of 30 days with the following results:

5-11-64.XLS

Relating time to invoices processed.

Day	Invoices Processed	Completion Time (hours)	Day	Invoices Processed	Completion Time (hours)
1	149	2.1	16	169	2.5
2	60	1.8	17	190	2.9
3	188	2.3	18	233	3.4
4	19	0.3	19	289	4.1
5	201	2.7	20	45	1.2
6	58	1.0	21	193	2.5
7	77	1.7	22	70	1.8
8	222	3.1	23	241	3.8
9	181	2.8	24	103	1.5
10	30	1.0	25	163	2.8
11	110	1.5	26	120	2.5
12	83	1.2	27	201	3.3
13	60	0.8	28	135	2.0
14	25	0.4	29	80	1.7
15	173	2.0	30	29	0.5

Hint: Determine which are the independent and dependent variables.
(a) Set up a scatter diagram.
(b) Assuming a linear relationship, use the least squares method to find the regression coefficients b_0 and b_1.
(c) Interpret the meaning of the Y intercept b_0 and the slope b_1 in this problem.
(d) Use the regression model developed in (b) to predict the amount of time it would take to process 150 invoices.
(e) Compute the standard error of the estimate.
(f) Compute the coefficient of determination r^2 and interpret its meaning.
(g) Compute the coefficient of correlation.
(h) Compute the Durbin-Watson statistic and, at the .05 level of significance, determine whether there is any autocorrelation in the residuals.
(i) Based on the results of (h), what conclusions can you reach concerning the validity of the model fit in (b)?
(j) At the .05 level of significance, is there evidence of a linear relationship between the amount of time and the number of invoices processed?
(k) Set up a 95% confidence interval estimate of the average amount of time taken to process 150 invoices.
(l) Set up a 95% prediction interval estimate of the amount of time it would take to process 150 invoices on a particular day.
(m) Perform a residual analysis on your results and determine the adequacy of the fit of the model.

11.65 Crazy Dave, a well-known baseball analyst, would like to study various team statistics for the 1995 baseball season to determine which variables might be useful in predicting the number of wins achieved by teams during the season. He has decided to begin by using the team earned run average (E.R.A.) to predict the number of wins. The data for the 28 major league teams are as shown on page 596:

BB95.TXT

Relating wins to E.R.A.

	American League			National League		
Team	Wins	E.R.A.		Team	Wins	E.R.A.
Boston	86	4.39		Florida	67	4.27
Cleveland	100	3.83		Cincinnati	85	4.03
Kansas City	70	4.49		Chicago Cubs	73	4.13
Minnesota	56	5.76		San Francisco	67	4.86
Toronto	56	4.88		Los Angeles	78	3.66
California	78	4.49		Pittsburgh	58	4.70
Seattle	78	4.52		San Diego	70	4.13
Texas	74	4.66		New York Mets	69	3.88
Detroit	60	5.49		St. Louis	62	4.09
Chicago White Sox	68	4.85		Philadelphia	69	4.21
Milwaukee	65	4.82		Atlanta	90	3.44
Oakland	67	4.93		Montreal	66	4.11
Baltimore	71	4.31		Houston	76	4.06
New York Yankees	79	4.56		Colorado	77	4.97

Note: Each team played 144 games in the 1995 season.

Hint: Determine which are the independent and dependent variables.

(a) Set up a scatter diagram.
(b) Assuming a linear relationship, use the least squares method to find the regression coefficients b_0 and b_1.
(c) Interpret the meaning of the Y intercept b_0 and the slope b_1 in this problem.
(d) Use the regression model developed in (b) to predict the number of wins for a team with an E.R.A. of 4.00.
(e) Compute the standard error of the estimate.
(f) Compute the coefficient of determination r^2 and interpret its meaning.
(g) Compute the coefficient of correlation.
(h) At the .05 level of significance, is there evidence of a linear relationship between the number of wins and the E.R.A.?
(i) Set up a 95% confidence interval estimate of the average number of wins for a team with an E.R.A. of 4.00.
(j) Set up a 95% prediction interval estimate of the number of wins for an individual team that has an E.R.A. of 4.00.
(k) Set up a 95% confidence interval estimate of the slope.
(l) Perform a residual analysis on your results and determine the adequacy of the fit of the model.
(m) The 28 teams constitute a population. In order to use statistical inference [as in (h)–(k)], the data must be assumed to represent a random sample. What "population" would this sample be drawing conclusions about?

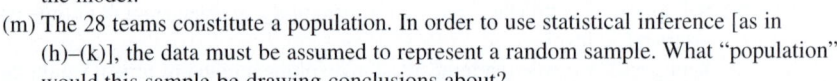

Case Study E—Predicting Sunday Newspaper Circulation

You are employed in the marketing department of a large nationwide newspaper chain. The parent company is interested in investigating the feasibility of beginning a Sunday edition for some of its newspapers. However, before proceeding with a final decision, it needs to estimate the amount of Sunday circulation that would be expected. In particular, it wishes to predict the Sunday circulation that would be obtained by newspapers (in three different cities) that have daily circulations of 200,000, 400,000, and 600,000, respectively.

You have been asked to develop a model that would enable you to make a prediction of the expected Sunday circulation and to write a report that presents your results and summarizes your findings. Toward this end, data collected from a sample of 35 newspapers are as follows:

 NEWSCIRC.TXT

Paper	Circulation (in 000)	
	Sunday	Daily
Des Moines Register	344.522	206.204
Philadelphia Inquirer	982.663	515.523
Tampa Tribune	408.343	321.626
New York Times	1,762.015	1,209.225
New York News	983.240	781.796
Sacramento Bee	338.355	273.844
Los Angeles Times	1,531.527	1,164.388
Boston Globe	798.298	516.981
Cincinnati Enquirer	348.744	198.832
Orange Co. Register	407.760	354.843
Miami Herald	553.479	444.581
Chicago Tribune	1,133.249	733.775
Detroit News	1,215.149	481.766
Houston Chronicle	620.752	449.755
Kansas City Star	423.305	288.571
Omaha World Herald	284.611	223.748
Denver Post	417.779	252.624
St. Louis Post-Dispatch	585.681	391.286
Portland Oregonian	440.923	337.672
Washington Post	1,165.567	838.902
Long Island Newsday	960.308	825.512
San Francisco Chronicle	704.322	570.364
Chicago Sun Times	559.093	537.780
Minneapolis Star Tribune	685.975	412.871
Baltimore Sun	488.506	391.952
Pittsburgh Press	557.000	220.465
Rocky Mountain News	432.502	374.009
Boston Herald	235.084	355.628
New Orleans Times-Picayune	324.241	272.280
Charlotte Observer	299.451	238.555
Hartford Courant	323.084	231.177
Rochester Democrat and Chronicle	262.048	133.239
St. Paul Pioneer Press	267.781	201.860
Providence Journal-Bulletin	268.060	197.120
L.A. Daily News	202.614	185.736

Source: From *Gale Directory of Publications*: 1994, 126th edition. Edited by Donald P. Boyden and John Krol. Gale Research, 1994 Copyright © 1994 by Gale Research, Inc. Reprinted by permission of the publisher.

Endnotes

1. In Section 12.12, we shall investigate multiple regression models in which at least one of the independent variables is categorical (see dummy variable models). Regression models in which the dependent variable is categorical involve the use of logistic regression (see References 3 and 10).

2. It is interesting and instructive to note that had we constructed the residual plots using the independent variable as the X-axis (instead of the estimated values $\hat{Y}$), the same conclusions would prevail.

3. Methods that analyze the impact of individual observations on the regression model are part of *influence analysis* (see References 1, 3, 4, 5, 8, 9, and 15).

References

1. Andrews, D. F., and D. Pregibon, "Finding the Outliers that Matter," *Journal of the Royal Statistical Society*, Ser. B., 1978, Vol. 40, pp. 85–93.

2. Anscombe, F. J., "Graphs in Statistical Analysis," *American Statistician*, 1973, Vol. 27, pp. 17–21.

3. Berenson, M. L. and D. M. Levine, *Basic Business Statistics: Concepts and Applications*, 6th ed. (Englewood Cliffs, NJ: Prentice-Hall, 1996).

4. Belsley, D. A., E. Kuh, and R. Welsch, *Regression Diagnostics: Identifying Influential Data and Sources of Collinearity* (New York: John Wiley, 1980).

5. Cook, R. D., and S. Weisberg, *Residuals and Influence in Regression* (New York: Chapman and Hall, 1982).

6. Conover, W. J., *Practical Nonparametric Statistics*, 2d ed. (New York: John Wiley, 1980).

7. Draper, N. R., and H. Smith, *Applied Regression Analysis*, 2d ed. (New York: John Wiley, 1981).

8. Hoaglin, D.C., and R. Welsch, "The Hat Matrix in Regression and ANOVA," *The American Statistician*, 1978, Vol. 32, pp. 17–22.

9. Hocking, R. R., "Developments in Linear Regression Methodology: 1959–1982," *Technometrics*, 1983, Vol. 25, pp. 219–250.

10. Hosmer, D., and S. Lemeshow, *Applied Logistic Regression* (New York: John Wiley, 1989).

11. *Microsoft Excel Version 7* (Redmond, WA: Microsoft Corp., 1996).

12. Neter, J., M. H. Kutner, C. J. Nachtsheim, and W. Wasserman, *Applied Linear Statistical Models*, 4th ed (Homewood, IL: Richard D. Irwin, 1996).

13. Ramsey, P. P., and P. H. Ramsey, "Simple Tests of Normality in Small Samples," *Journal of Quality Technology*, 1990, Vol. 22, pp. 299–309.

14. Tukey, J.W., "Data Analysis, Computation and Mathematics," *Quarterly Journal of Applied Mathematics*, 1972, Vol. 30, pp. 51–65.

15. Velleman, P. F., and R. Welsch, "Efficient Computing of Regression Diagnostics," *The American Statistician*, 1981, Vol. 35, pp. 234–242.

License Agreement

YOU SHOULD CAREFULLY READ THE FOLLOWING TERMS AND CONDITIONS BEFORE BREAKING THE SEAL ON THE PACKAGE. AMONG OTHER THINGS, THIS AGREEMENT LICENSES THE ENCLOSED SOFTWARE TO YOU AND CONTAINS WARRANTY AND LIABILITY DISCLAIMERS. BY BREAKING THE SEAL ON THE PACKAGE, YOU ARE ACCEPTING AND AGREEING TO THE TERMS AND CONDITIONS OF THIS AGREEMENT. IF YOU DO NOT AGREE TO THE TERMS OF THIS AGREEMENT, DO NOT BREAK THE SEAL. YOU SHOULD PROMPTLY RETURN THE PACKAGE UNOPENED.

LICENSE

Subject to the provisions contained herein, Prentice-Hall, Inc. ("PH") hereby grants to you a non-exclusive, non-transferable license to use the object code version of the computer software product ('Software') contained in the package on a single computer of the type identified on the package.

SOFTWARE AND DOCUMENTATION

PH shall furnish the Software to you on media in machine-readable object code form and may also provide the standard documentation (Documentation') containing instructions for operation and use of the Software.

LICENSE TERM AND CHARGES

The term of this license commences upon delivery of the Software to you and is perpetual unless earlier terminated upon default or as otherwise set forth here

TITLE

itle, and ownership right, and intellectual property rights in and to the Software and Documentation shall remain in PH and/or suppliers to PH of programs contained in the Software. The Software is provided for your own internal use under this license. This license does not include the right to sublicense and is personal to you and therefore may not be assigned (by operation of law or otherwise) or transferred without the prior written consent of PH. You acknowledge that the Software in source code form remains a confidential trade secret of PH and/or its suppliers and therefore you agree not to attempt to decipher or decompile, modify, disassemble, reverse engineer or prepare derivative works of the Software or develop source code for the Software or knowingly allow others to do so. Further, you may not copy the Documentation or other written materials accompanying the Software.

UPDATES

This license does not grant you any right, license, or interest in and to any improvements, modifications, enhancements, or updates to the Software and Documentation. Updates, if available, may be obtained by you at PH's then current standard pricing, terms, and conditions.

LIMITED WARRANTY AND DISCLAIMER

PH warrants that the media containing the Software, if provided by PH, is free from defects in material and workmanship under normal use for a period of sixty (60) days from the date you purchased a license to use it.

THIS IS A LIMITED WARRANTY AND IT IS THE ONLY WARRANTY MADE BY PH. THE SOFTWARE IS PROVIDED 'AS IS' AND PH SPECIFICALLY DISCLAIMS ALL WARRANTIES OF ANY KIND, EITHER EXPRESS OR IMPLIED, INCLUDING, BUT NOT LIMITED TO, THE IMPLIED WARRANTY OF MERCHANTABILITY AND FITNESS FOR A PARTICULAR PURPOSE. FURTHER, COMPANY DOES NOT WARRANT, GUARANTY, OR MAKE ANY REPRESENTATIONS REGARDING THE USE, OR THE RESULTS OF THE USE, OF THE SOFTWARE IN TERMS OF CORRECTNESS, ACCURACY, RELIABILITY, CURRENTNESS, OR OTHERWISE AND DOES NOT WARRANT THAT THE OPERATION OF ANY SOFTWARE WILL BE UNINTERRUPTED OR ERROR FREE. COMPANY EXPRESSLY DISCLAIMS ANY WARRANTIES NOT STATED HEREIN. NO ORAL OR WRITTEN INFORMATION OR ADVICE GIVEN BY PH, OR ANY PH DEALER, AGENT, EMPLOYEE OR OTHERS SHALL CREATE, MODIFY OR EXTEND A WARRANTY OR IN ANY WAY INCREASE THE SCOPE OF THE FOREGOING WARRANTY, AND NEITHER SUBLICENSEE OR PURCHASER MAY RELY ON ANY SUCH INFORMATION OR ADVICE. If the media is subjected to accident, abuse, or improper use; or if you violate the terms of this Agreement, then this warranty shall immediately be terminated. This warranty shall not apply if the Software is used on or in conjunction with hardware or programs other than the unmodified version of hardware and programs with which the Software was designed to be used as described in the Documentation.

LIMITATION OF LIABILITY

Your sole and exclusive remedies for any damage or loss in any way connected with the Software are set forth below. UNDER NO CIRCUMSTANCES AND UNDER NO LEGAL THEORY, TORT, CONTRACT, OR OTHERWISE, SHALL PH BE LIABLE TO YOU OR ANY OTHER PERSON FOR ANY INDIRECT, SPECIAL, INCIDENTAL, OR CONSEQUENTIAL DAMAGES OF ANY CHARACTER INCLUDING, WITHOUT LIMITATION, DAMAGES FOR LOSS OF GOODWILL, LOSS OF PROFIT, WORK STOPPAGE, COMPUTER FAILURE OR MALFUNCTION, OR ANY AND ALL OTHER COMMERCIAL DAMAGES OR LOSSES, OR FOR ANY OTHER DAMAGES EVEN IF PH SHALL HAVE BEEN INFORMED OF THE POSSIBILITY OF SUCH DAMAGES, OR FOR ANY CLAIM BY ANY OTHER PARTY. PH'S THIRD PARTY PROGRAM SUPPLIERS MAKE NO WARRANTY, AND HAVE NO LIABILITY WHATSOEVER, TO YOU. PH's sole and exclusive obligation and liability and your exclusive remedy shall be: upon PH's election (i) the replacement of your defective media; or (ii) the repair or correction of your defective media if PH is able, so that it will conform to the above warranty; or (iii) if PH is unable to replace or repair, you may terminate this license by returning the Software. Only if you inform PH of your problem during the applicable warranty period will PH be obligated to honor this warranty. You may contact PH to inform PH of the problem as follows:

SOME STATES OR JURISDICTIONS DO NOT ALLOW THE EXCLUSION OF IMPLIED WARRANTIES OR LIMITATION OR EXCLUSION OF CONSEQUENTIAL DAMAGES, SO THE ABOVE LIMITATIONS OR EXCLUSIONS MAY NOT APPLY TO YOU. THIS WARRANTY GIVES YOU SPECIFIC LEGAL RIGHTS AND YOU MAY ALSO HAVE OTHER RIGHTS WHICH VARY BY STATE OR JURISDICTION.

MISCELLANEOUS

If any provision of this Agreement is held to be ineffective, unenforceable, or illegal under certain circumstances for any reason, such decision shall not affect the validity or enforceability (i) of such provision under other circumstances or (ii) of the remaining provisions hereof under all circumstances and such provision shall be reformed to and only to the extent necessary to make it effective, enforceable, and legal under such circumstances. All headings are solely for convenience and shall not be considered in interpreting this Agreement. This Agreement shall be governed by and construed under New York law as such law applies to agreements between New York residents entered into and to be performed entirely within New York, except as required by U.S. Government rules and regulations to be governed by Federal law.

YOU ACKNOWLEDGE THAT YOU HAVE READ THIS AGREEMENT, UNDERSTAND IT, AND AGREE TO BE BOUND BY ITS TERMS AND CONDITIONS. YOU FURTHER AGREE THAT IT IS THE COMPLETE AND EXCLUSIVE STATEMENT OF THE AGREEMENT BETWEEN US THAT SUPERSEDES ANY PROPOSAL OR PRIOR AGREEMENT, ORAL OR WRITTEN, AND ANY OTHER COMMUNICATIONS BETWEEN US RELATING TO THE SUBJECT MATTER OF THIS AGREEMENT.

U.S. GOVERNMENT RESTRICTED RIGHTS

Use, duplication or disclosure by the Government is subject to restrictions set forth by the subparagraphs (a) through (d) of the Commercial Computer Restricted Rights clause at FAR 52.227-19 when applicable, or in subparagraph (c)(l)(ii) of the Rights in Technical Data and Computer Software clause at DFARS 252.227-7013, and in similar clauses in the NASA FAR Supplement.